Hassan Errami

Efficient Methods for Symbolic Analysis of Complex Reaction Networks

Hassan Errami

Efficient Methods for Symbolic Analysis of Complex Reaction Networks

Südwestdeutscher Verlag für Hochschulschriften

Impressum / Imprint

Bibliografische Information der Deutschen Nationalbibliothek: Die Deutsche Nationalbibliothek verzeichnet diese Publikation in der Deutschen Nationalbibliografie; detaillierte bibliografische Daten sind im Internet über http://dnb.d-nb.de abrufbar.
Alle in diesem Buch genannten Marken und Produktnamen unterliegen warenzeichen-, marken- oder patentrechtlichem Schutz bzw. sind Warenzeichen oder eingetragene Warenzeichen der jeweiligen Inhaber. Die Wiedergabe von Marken, Produktnamen, Gebrauchsnamen, Handelsnamen, Warenbezeichnungen u.s.w. in diesem Werk berechtigt auch ohne besondere Kennzeichnung nicht zu der Annahme, dass solche Namen im Sinne der Warenzeichen- und Markenschutzgesetzgebung als frei zu betrachten wären und daher von jedermann benutzt werden dürften.

Bibliographic information published by the Deutsche Nationalbibliothek: The Deutsche Nationalbibliothek lists this publication in the Deutsche Nationalbibliografie; detailed bibliographic data are available in the Internet at http://dnb.d-nb.de.
Any brand names and product names mentioned in this book are subject to trademark, brand or patent protection and are trademarks or registered trademarks of their respective holders. The use of brand names, product names, common names, trade names, product descriptions etc. even without a particular marking in this works is in no way to be construed to mean that such names may be regarded as unrestricted in respect of trademark and brand protection legislation and could thus be used by anyone.

Coverbild / Cover image: www.ingimage.com

Verlag / Publisher:
Südwestdeutscher Verlag für Hochschulschriften
ist ein Imprint der / is a trademark of
OmniScriptum GmbH & Co. KG
Heinrich-Böcking-Str. 6-8, 66121 Saarbrücken, Deutschland / Germany
Email: info@svh-verlag.de

Herstellung: siehe letzte Seite /
Printed at: see last page
ISBN: 978-3-8381-3970-8

Zugl. / Approved by: Kassel, University of Kassel, Diss., 2013

Copyright © 2014 OmniScriptum GmbH & Co. KG
Alle Rechte vorbehalten. / All rights reserved. Saarbrücken 2014

Efficient Methods for Symbolic Analysis of Complex Reaction Networks

by

Hassan Errami

To my wife Ivonne Diana, my children, and my parents

"Everything Should Be Made as Simple as Possible, But Not Simpler."

Albert Einstein

Abstract

The identification of chemical mechanism that can exhibit oscillatory phenomena in reaction networks are currently of intense interest. In particular, the parametric question of the existence of Hopf bifurcations has gained increasing popularity due to its relation to the oscillatory behavior around the fixed points. However, the detection of oscillations in high-dimensional systems and systems with constraints by the available symbolic methods has proven to be difficult. The development of new efficient methods are therefore required to tackle the complexity caused by the high-dimensionality and non-linearity of these systems.

In this work, we mainly present efficient algorithmic methods to detect Hopf bifurcation fixed points in (bio)-chemical reaction networks with symbolic rate constants, thereby yielding information about their oscillatory behavior of the networks. The methods use the representations of the systems on convex coordinates that arise from stoichiometric network analysis. One of the methods called *HoCoQ* reduces the problem of determining the existence of Hopf bifurcation fixed points to a first-order formula over the ordered field of the reals that can then be solved using computational-logic packages. The second method called *HoCaT* uses ideas from tropical geometry to formulate a more efficient method that is incomplete in theory but worked very well for the attempted high-dimensional models involving more than 20 chemical species.

The instability of reaction networks may lead to the oscillatory behaviour. Therefore, we investigate some criterions for their stability using convex coordinates and quantifier elimination techniques.

We also study Muldowney's extension of the classical Bendixson-Dulac criterion for excluding periodic orbits to higher dimensions for polynomial vector fields and we discuss the use of simple conservation constraints and the use of parametric constraints for describing simple convex polytopes on which periodic orbits can be excluded by Muldowney's criteria.

All developed algorithms have been integrated into a common software framework called *PoCaB* (**p**latform t**o** explore bi**o**-**c**hemical reaction networks by **a**lgebraic methods) allowing for automated computation workflows from the problem descriptions. *PoCaB* also contains a database for the algebraic entities computed from the models of chemical reaction networks.

Contents

Abstract	iii
List of Figures	vii
List of Tables	viii
Abbreviations	x
Symbols	xi

1	**Introduction**	**1**
2	**Fundamentals**	**6**
2.1	Chemical Reaction Networks	6
	2.1.1 Stoichiometric Network Analysis	7
	2.1.2 Modeling Chemical Systems by Pseudolinear Ordinary Differential Equations	8
2.2	Quantifier Elimination and Formula Simplification	9
2.3	Linear Programming	11
3	**Generation of Algebraic Data**	**13**
3.1	Computation of Basic Algebraic Data	13
3.2	Flux Cone and Extreme Currents	15
3.3	Computation of Jacobian Matrix Using Convex Coordinates	16
3.4	Algebraic Data for Graph-Theoretic Representation of the Reaction Systems	17
3.5	Deficiency Value of the Reaction Network	18
3.6	*PoCaB*: A Software Infrastructure to Explore Algebraic Methods for (Bio)-Chemical Reaction Networks	19
	3.6.1 Representation of Reaction Networks	19
	3.6.2 Database of Algebraic Entities	20
	3.6.2.1 Data Source	20
	3.6.2.2 Software Workflow and Components	20
	3.6.2.3 Content of Database	22
	3.6.2.4 Statistical Summary	22

4 Detection of Hopf Bifurcations Using Convex Coordinates   25
   4.1   Hopf Bifurcations and Invariant Manifolds 25
        4.1.1   Conditions for Existence of Hopf Bifurcations 25
        4.1.2   Reduction to Invariant Manifolds 27
        4.1.3   Stability and Bifurcations for Semi-Explicit DAEs 28
   4.2   *HoCoQ*: An Algorithm for Computing Hopf Bifurcations using Convex Coordinates and Quantifier Elimination 31
        4.2.1   Pre-processing 32
        4.2.2   Polyhedral Computations 32
        4.2.3   Computation of the Hopf Condition in Convex Coordinates 32
               4.2.3.1   Computation of the Jacobian in Reaction Space 33
               4.2.3.2   Jacobian on the Reduced Manifold 33
               4.2.3.3   Semi-Algebraic Description of Hopf Bifurcations 33
        4.2.4   Integration of Computational Logic Tools 33
        4.2.5   Pseudo-Code of the *HoCoQ* Algorithm 34
        4.2.6   Computation of Examples using *HoCoQ* Method 35
               4.2.6.1   Phosphofructokinase Reaction 35
               4.2.6.2   Enzymatic Transfer of Calcium Ions 37
               4.2.6.3   Model of Calcium Oscillations in the Cilia of Olfactory Sensory Neurons 38
   4.3   *HoCaT*: Algorithm for Computing Hopf Bifurcations using Convex Coordinates and Tropical Geometry 40
        4.3.1   Sufficient Conditions for a Positive Solution of a Single Multivariate Polynomial Equation 41
        4.3.2   Summarizing the *HoCaT* Algorithm 46
        4.3.3   Computation of Examples Using the *HoCaT* Method 46
               4.3.3.1   Phosphofructokinase Reaction 46
               4.3.3.2   Enzymatic Transfer of Calcium Ions 47
               4.3.3.3   Model of Calcium Oscillations in the Cilia of Olfactory Sensory Neurons 48
               4.3.3.4   Electro-Oxidation of Methanol 49
               4.3.3.5   Methylene Blue Oscillator System 50
               4.3.3.6   Mitogen-Activated Protein Kinase ($MAPK$) 52
               4.3.3.7   Models of Genetic Circuits 54
               4.3.3.8   Control of DNA Replication in Fission Yeast 57

5 On Muldowney's Criteria for Polynomial Vector Fields with Constraints   58
   5.1   Introduction and Preliminaries 58
        5.1.1   The Bendixson-Dulac Criterion for 2-Dimensional Vector Fields . . 59
        5.1.2   Muldowney's Extensions of the Bendixson-Dulac Criterion to Higher Dimensions 60
               5.1.2.1   Extending Muldowney's Criteria with Dulac Functions. . 61
               5.1.2.2   Using Conservation Constraints. 61
               5.1.2.3   Parametric Specification of a Convex Subset. 62
   5.2   Case Studies 62
        5.2.1   The SIRS Epidemiological Model 62

		5.2.1.1	Using Ad-hoc Reductions to 2D-Models	63
		5.2.1.2	Computations on the 3D-Model	64
	5.2.2		A Model of Viral Dynamics	64

6 Computing Stability in Convex Coordinates   67
    6.1 Computing Stability Using Hurwitz Criterion 67
    6.2 Computing Stability Using Gantmacher-Stieltjes Criterion 71
    6.3 Computation of Mixing Stability 73

7 Summary and Outlook   76

Bibliography   79

List of Figures

3.1	Extreme currents of flux cone	15
3.2	Components of SBML file	20
4.1	$HoCoQ$ method	31
4.2	$HoCaT$ method	41
4.3	Newton polytope and a separating hyperplane	43
4.4	A gene regulated by a polymer of its protein	55
5.1	The 2D- and 3D-Tuckwell-Wan examples	64

List of Tables

3.1 Summary of results in Biomodels and KEGG database 24

4.1 Computation of Hopf bifurcations in the phosphofructokinase reaction using $HoCoQ$ algorithm 36
4.2 Computation of Hopf bifurcations in the model "enzymatic transfer of calcium ions" using $HoCoQ$ algorithm 38
4.3 Computation of Hopf bifurcations in model of calcium Oscillations using $HoCoQ$ algorithm 40
4.4 Computation of Hopf bifurcations in the phosphofructokinase reaction using $HoCaT$ algorithm 48
4.5 Computation of Hopf bifurcations in the model "enzymatic transfer of calcium ions" using $HoCaT$ algorithm 48
4.6 Computation of Hopf bifurcations in the model "calcium oscillations in the cilia of olfactory sensory neurons" using $HoCaT$ algorithm 48
4.7 Computation of Hopf bifurcations in the model "electro-oxidation of methanol" using $HoCaT$ algorithm 50
4.8 Results of the computation of Hopf bifurcations in 1-face and 2- faces using $HoCaT$ 52

5.1 Results for the 2D-Tuckwall-Wan example (cf. Fig. 5.1) on the full positive octant 65

6.1 Computation of stability in the phosphofructokinase reaction in convex coordinates using Hurwitz condition 69
6.2 Computation of stability in the model "Enzymatic transfer of calcium ions" in convex coordinates using Hurwitz condition 69
6.3 Computation of stability in the model "calcium oscillations in the cilia of olfactory sensory neurons" using convex coordinates and Hurwitz condition 70
6.4 Computation of stability in the model "electro-oxidation of methanol" using convex coordinates and Hurwitz condition 70
6.5 Computation of stability in the phosphofructokinase reaction using convex coordinates and Gantmacher-Stieltjes condition 72
6.6 Computation of stability in the model "enzymatic transfer of calcium ions" using convex coordinates and Gantmacher-Stieltjes condition 72
6.7 Computation of stability in the model "calcium oscillations in the cilia of olfactory sensory neurons" using convex coordinates and Gantmacher-Stieltjes condition 73
6.8 Computation of stability in the model "electro-oxidation of methanol" using convex coordinates and Gantmacher-Stieltjes condition 73
6.9 Computation of mixing stability in the phosphofructokinase reaction . . . 74

6.10 Computation of mixing stability in the model "enzymatic transfer of calcium ions" 74
6.11 Computation of mixing stability in the model "calcium oscillations in the cilia of olfactory sensory neurons" 75
6.12 Computation of mixing stability in the model "electro-oxidation of methanol" 75

Abbreviations

ATP	Adenosine Tri-Phosphate
CAD	Cylindrical Algebraic Decomposition
DAE	Differential Algebraic Equation
KEGG	Kyoto Encyclopedia of Genes and Genomes
KGML	KEGG Markup Language
LP	Linear Programming
MAPK	Mitogen-Activated Protein Kinase
MBO	Methylen Blue Oscillator
MILP	Mixed-Integer Linear Programming
MIQCP	Mixed-Integer Quadratically Constrained Programming
MIQP	Mixed-Integer Quadratic Programming
ODE	Ordinary Differential Equation
QCP	Quadratically Constrained Programming
QP	Quadratic Programming
SBML	System Biology Markup Language
SIRS	Systemic Inflammatory Response Syndrome
SMT	Satisfiability Modulo Theories
SNA	Stoichiometric Network Analysis

Symbols

$\mathcal{S}$	Stoichiometric matrix
$\mathcal{K}$	Kinetic matrix
$v(x,k)$	Flux vector
$\mathcal{E}$	Set of extreme currents
j_i	Convex parameter
x_i	Concentration of the ith-species
k_i	Rate constant of the ith-reaction
I_a, I_k	Incidence matrices
Y	Adjacency matrix
$I_k \Psi(x)$	Vector of monomials
$\mathcal{N}$	(Bio)-Chemical network
Jac	Jacobian matrix
$\widehat{\text{Jac}}$	Convex Jacobian matrix
diag	Diagonal matrix
Δ_i	The ith-minor of the Hurwitz matrix
χ_i	Characteristic polynomial
newton(g)	Newton polytope of g
frame(g)	Set of exponent vectors of g

Chapter 1

Introduction

The dynamics of (bio)-chemical systems are usually described by power-law kinetics, i.e. the reaction rates are proportional to some power of the species concentrations involved. If it is assumed that these (bio)-chemical systems follow mass action kinetics then the dynamics of these reactions can be represented by ordinary differential equations (ODEs) for systems without additional constraints or by differential algebraic equations (DAEs) for systems with constraints.

An important task of the chemist is the identification of chemical mechanism that can exhibit exotic phenomena caused by instability of the steady state, such as switching between multiple steady states, explosions, oscillations, or even more complicated phenomena, which are known by mathematicians as motion on an attracting k-torus and strange attractors (chaos) [12]. In particular, the oscillatory phenomena in chemistry are currently of intense interest.

Stability analysis is by far the most effective technique because it uses linearized equations of motion; the direct identification of local and global oscillations requires a tedious analysis of the complete nonlinear equations of motion, which is possible only for simple systems. In principle, oscillations and various exotic dynamics may occur in systems of differential equations even when no steady states are unstable. Although stability analysis is by far the most effective method, it is not a direct test for the existence of exotic dynamics; however the differential equations for chemical systems appear to be special. So far, no remotely plausible model has been constructed of a chemical network with oscillations and only stable steady states. Hence linear stability analysis may be used with some confidence to determine which chemical models are capable of exotic dynamics in cases that are far too complex for the direct identification of oscillations [12].

Chapter 1. *Introduction*

The development of symbolic methods for studying local and global oscillations was a topic of considerable research effort in the last decade. Especially, the parametric question of the existence of Hopf bifurcations attracts more interests due to its relation to oscillatory behavior around the fixed points. A Hopf bifurcation is a local bifurcation in which the equilibrium point of a dynamical system loses stability when a pair of complex conjugate eigenvalues of the linearization around the equilibrium point cross the imaginary axis of the complex plan. Some low dimensional reaction systems without additional constraints have been already investigated [4, 5, 30]. The parametric question for a parameterized polynomial vector field whether fixed points undergo a Hopf bifurcation is not only known to be decidable but also lies within the realm of semi-algebraic sets [23, 30, 52]. Hence, the questions whether there are Hopf bifurcation fixed points inducing local oscillations can be reduced to decidable questions on semi-algebraic sets for polynomial vector fields via the well-known Routh-Hurwitz criterion [39].

A fully algebraic method for the computation of Hopf bifurcation fixed points for systems with polynomial vector fields has already been introduced by El Kahoui and Weber [23] using the powerful technique of quantifier elimination on real closed fields [83]. This technique has already been successfully applied to the mass action kinetics of low dimensions [80]. However deciding its occurrence in high dimensional differential equations and in differential algebraic equations has proven to be difficult in practice. Nevertheless, quite a few conclusions regarding the dynamics can be drawn from the structure of the reaction network itself. In this context, there has been a surge in the development of algebraic methods that are based on the structure of the network and the associated stoichiometry of the chemical species. These methods are aimed at understanding the qualitative behavior of the network. Using ideas from so called *stoichiometric network analysis* (SNA) [12] it is possible to analyze the system dynamics in flux space instead of the concentration space and to represent the space of the steady states with a combination of subnetworks using methods from convex geometry. Methods for detecting Hopf bifurcations using similar approaches have been used in several "hand computations" in a semi-algorithmic way for parametric systems, the most elaborate of which is described in [30].

In the study of differential equations the analysis of global oscillations is seen as an important goal in addition to describing the dynamics around fixed points. However, already for two-dimensional polynomial systems the global question whether there are periodic solutions (and thus oscillations) is related to Hilbert's 16th problem, which is still unsolved [42]. For the two-dimensional case the Bendixson-Dulac criterion gives a sufficient condition for the non-existence of periodic orbits. This criterion is parameterized by a Dulac function, and various techniques have been proposed to construct

Dulac functions, which range from algebraic constructions for special systems to techniques involving the solution of certain partial differential equations [8–11, 55]. For the higher-dimensional case there are extensions of the criterion of Bendixson-Dulac that also allow the use of Dulac functions [51]. However, little work seems to have been done to construct Dulac functions in the higher dimensional cases, except for addressing it as a problem [92, 93]. Most of these extensions involve first integrals of the various field [24, 85, 88] and thus might leave the realm of computations on semi-algebraic sets, but the criteria of Muldowney [51] (parameterised by a matrix norm) lead to quantifier elimination problems over real closed fields (at least for the L^1- and L^∞-norm). Moreover, the common case of algebraic constraints in the simple form of conservation constraints have been used in ad hoc form by many authors, mainly to reduce 3D systems to 2D systems in order to be able to use the classical Bendixson-Dulac criterion, but have not been discussed in a more general setting.

As already mentioned above, chemical and biological systems with constraints can be modeled by differential algebraic equations(DAEs). The analytical theory of these equations has not yet achieved the same level as the theory of normal systems; this concerns in particular the analysis of singularities and bifurcations (although both [61] and [65] contain a chapter on singularities). Hopf bifurcation of semi-explicit DAEs are studied by Venkatasubramanian et al. [89]. Rabier [60] discusses the case of quasi-linear DAEs; Beardmore and Webster [1] analyse a certain kind of singular quasi-linear DAEs. Different forms of bifurcations have been considered by von Sosen [90]. However, one should note that most of these works treat fairly special situations; general non-linear DAEs have not been considered so far. Compared to normal systems, already the usual existence and uniqueness theory of DAEs is much more complicated due to the possible existence of hidden integrability conditions. In fact, much of the above mentioned literature on the analytical theory of DAEs is concerned with this problem and its effects on the numerical analysis.

The main goal of this work is the investigation and development of algorithmic methods for detection of oscillations in biological and chemical systems with particular emphasis on questions concerning the occurrence of Hopf bifurcation fixed points. The new developed approaches improve and combine the ideas and techniques listed above. Another major goal is the integration of all developed algorithms into a common software framework. A first step of the analysis of chemical systems using algebraic methods is the generation of algebraic data that describe their reaction laws. Therefore, in Chapter 3 the approaches for generating algebraic data from (bio)-chemical networks will be discussed, and a developed software infrastructure to explore algebraic methods for (bio)-chemical reaction networks will be presented. We discuss also in this chapter the computation of extreme currents and thus the transformation of original coordinates to the convex coordinates.

In Chapter 4, we present efficient algorithmic methods to detect Hopf bifurcations in complex (bio)-chemical reaction networks with symbolic rate constant; In the first algorithmic method we applied a combination of the known (and already demonstrated) algorithmic reduction to quantifier elimination problems over the reals and the algorithmic solutions of these problems with techniques arising from stoichiometric network analysis, such as the use of convex coordinates. Technically this combination will yield an existentially quantified problem that consists of determining Hopf-bifurcation fixed point with empty unstable manifold involving the conjunction of the following condition: an equality condition on the principal minor $\Delta_{n-1} = 0$ of the Jacobian of the vector field in conjunction with inequality conditions on $\Delta_{n-2} > 0 \wedge \cdots \wedge \Delta_1 > 0$ and positivity conditions on the variables and parameters.

Another method for the parametric detection of Hopf bifurcations that also uses techniques of stoichiometric networks analysis is presented as the second algorithm. This algorithm is based on the basic observation that the condition for existence of Hopf bifurcation fixed points when using convex coordinates is given by the single polynomial equation $\Delta_{n-1} = 0$ (together with positivity conditions on the convex coordinates) and (drop resp. delaying a test for the existence of unstable empty manifolds on already determined witness points for Hopf bifurcations). Therefore the main algorithmic problem is to determine whether a single multivariate polynomial can have a zero for positive coordinates. For this purpose we provide heuristics on the basis of the Newton polytope that ensure the existence of positive and negative values of the polynomial for positive coordinates. We evaluate our methods on a variety of examples, some of which have a dimensions greater than 20. Considering the performance of our methods, we could now analyse medium sized networks in their unreduced forms, a task for which the only previously available was the analysis of quasi-steady state approximations.

In Chapter 5, we study Muldowney's extension of the classical Bendixson-Dulac criterion for excluding periodic orbits to higher dimensions for polynomial vector fields. Using the formulation of Muldowney's sufficient criteria for excluding periodic orbits of the parameterized vector field on a convex set as a quantifier elimination problem over the ordered field of the reals we provide case studies of some systems arising in the life sciences. We discuss the use of simple conservation constraints and the use of parametric constraints for describing simple convex polytopes on which periodic orbits can be excluded by Muldowney's criteria.

The instability of a (bio)-chemical system gives rise to the existence of an exotic dynamic. Hence, computing stability can be used to determine chemical networks and subnetworks

Chapter 1. *Introduction* 5

which are capable of oscillations. The computation of stability using stoichiometric network analysis and quantifier elimination is discussed in Chapter 6. Two criterion are thereby tested, namely Huwritz criterion and Gantmacher-Stieltjes criterion. We also applied and tested *mixing stability* which invented by Clarke [12] for computing stability of extreme subnetworks.

Chapter 2

Fundamentals

2.1 Chemical Reaction Networks

A chemical reaction occurs when two or more chemical species react to become new chemical species. This process is usually presented by an equation where the *reactants* are given on the left hand side of an arrow and the *products* on the right hand side, the numbers next to the species called *stoichiometric coefficients* present the amount to which a chemical species participates in a reaction and the parameter on the arrow called *rate constant* stands for an experimental constant influencing the reaction velocity. a chemical reaction is called *irreversible*, if it proceeds only in one direction, and is called *reversible*, if it proceeds in either directions. In order to be compatible with thermodynamics, in reversible reactions the difference between the kinetic exponents of the reverse and forward reaction must be equal to the stoichiometric coefficient for each species, this is referred to as mass action kinetics.

An example of a chemical reaction, as it usually appears in the literature, is the following:

$$A + B \xrightarrow{k} 3A + C$$

In this reaction, one unit of chemical *species* A and one of B react (at reaction rate k) to form three units of A and one of C. The concentrations of these three species, denoted by x_a, x_b and x_c, will change in time as the reaction occurs. Under the assumption of *mass-action kinetics*, species A and B react at a rate proportional to the product of their concentrations, where the proportionality constant is the rate constant k. Noting that the reaction yields a net change of two units in the amount of A [30, 57, 73], we obtain the following corresponding differential equations:

$$\frac{d}{dt}x_a = 2kx_ax_b$$
$$\frac{d}{dt}x_b = -kx_ax_b$$
$$\frac{d}{dt}x_c = kx_ax_b \qquad (2.1)$$

A *chemical reaction network* can be defined as a finite set of chemical reactions. It can presented as a finite directed graph whose vertices are labeled by complexes and whose edges are labeled by parameters(reaction rate constants). Specifically, the digraph is denoted $G = (V, E)$, with vertex set $V = \{1, 2, ..., m\}$ and edge set $E \subseteq \{(i, j) \in V \times V : i \neq j\}$. A network is reversible if the graph G is undirected, in which case each undirected edge has two labels k_{ij} and k_{ji} [57, 73].

2.1.1 Stoichiometric Network Analysis

The usual way to understand the behavior of mass-action chemical systems is to observe the time evolution of the species concentration. This can be mathematically represented by coupled differential equations, where each equation represent a change in a correspondent species concentration. Thus the analysis of the chemical systems in concentration space turned out to be hard by increasing number of chemical species.

Clarke has introduced in 1980 a new method called stoichiometric network analysis (SNA) to analyze the stability of mass-action chemical reaction systems[12]. The idea of SNA is to observe the dynamics of the system in the reaction space instead of concentration space. This leads to expand the steady state into a combination of subnetworks that form a convex cone in the flux-space called *flux cone*. Provided that all reactions are unidirectional or irreversible, the intersection of the null-space with the semipositive orthant of the flux space procedure results in a set of rays or edges starting at 0, which fully describe the cone. The edges are represented by vectors and any admissible steady state of the system is a positive combination of these vectors. From a biological perspective, these edges characterize important pathways of the metabolic network. In the case of a pointed cone, where 0 is a vertex, they connect inputs to outputs with a minimal set of reaction [91].

2.1.2 Modeling Chemical Systems by Pseudolinear Ordinary Differential Equations

The differential equations in chemical reaction networks usually are constrained reflecting various physical conservation laws. The systems with linear constraints often found in chemical reaction networks can easily be generalized to *pseudolinear ordinary differential equations*. The basic underlying property of the considered differential equations is captured by the following definition.

Definition 2.1. We call an autonomous system of ordinary differential equations $\dot{\mathbf{x}} = \boldsymbol{\phi}(\mathbf{x})$ for an unknown function $\mathbf{x} : \mathbb{R} \to \mathbb{R}^n$ *pseudolinear*, if its right hand side can be written in the form $\boldsymbol{\phi}(\mathbf{x}) = N\boldsymbol{\psi}(\mathbf{x})$ with a constant matrix $N \in \mathbb{R}^{n \times m}$ and some vector valued function $\boldsymbol{\psi} : \mathbb{R}^n \to \mathbb{R}^m$.

Obviously, any *polynomially* nonlinear system can be written in such a form, if we take for $\boldsymbol{\psi}(\mathbf{x})$ the vector of all terms appearing on the right hand side of the system. As one can see from the following two lemmata, the pseudolinear structure is of interest only in the case that the matrix N does not possess full row rank and hence the range of N is not the full space $\mathbb{R}^n$. In the sequel, we will always assume that the function $\boldsymbol{\psi}$ satisfies $m \geq n$, as this is usually the case in applications like reaction kinetics.

Lemma 2.2. *For a pseudolinear system* $\dot{\mathbf{x}} = N\boldsymbol{\psi}(\mathbf{x})$ *any affine subspace of the form* $\mathcal{A}_{\mathbf{y}} = \mathbf{y} + \operatorname{im} N \subseteq \mathbb{R}^n$ *for an arbitrary constant vector* $\mathbf{y} \in \mathbb{R}^n$ *defines an invariant manifold.*

Proof. Obviously, we have $\dot{\mathbf{x}}(t) \in \operatorname{im} N$ for all times t and $T_{\mathbf{x}}\mathcal{A}_{\mathbf{y}} = \operatorname{im} N$ for all points $\mathbf{x} \in \mathcal{A}_{\mathbf{y}}$ by definition of an affine space. Thus, if $\mathbf{x}(0) \in \mathcal{A}_{\mathbf{y}}$, then the whole trajectory will stay in $\mathcal{A}_{\mathbf{y}}$. □ □

Remark 2.3. For the application in reaction kinetics, the following minor strengthening of Lemma 2.2 is of interest. Assume that the function $\boldsymbol{\psi}$ satisfies additionally $\boldsymbol{\psi}(\mathbf{x}) \in \mathbb{R}^m_{\geq 0}$ for all $\mathbf{x} \in \mathbb{R}^n_{\geq 0}$ which is for example trivially the case when each component of $\boldsymbol{\psi}$ is a polynomial with positive coefficients. If we solve our differential equation for non-negative initial data $\mathbf{x}(0) = \mathbf{x}_0 \in \mathbb{R}^n_{\geq 0}$, then the solution always stays in the convex polyhedral cone $\mathbf{x}_0 + \left\{\sum_{i=1}^m \lambda_i \mathbf{n}_i \mid \forall i : \lambda_i \geq 0\right\}$ where the vectors $\mathbf{n}_i$ are the columns of the matrix N. Indeed, in this case the tangent vector $\dot{\mathbf{x}}(t)$ along the trajectory is trivially always a non-negative linear combination of the columns of N.

Lemma 2.4. *Let* $\mathbf{v}^T \cdot \mathbf{x} = \text{Const}$ *for some vector* $\mathbf{v} \in \mathbb{R}^n$ *be a linear conservation law of a pseudolinear system* $\dot{\mathbf{x}} = N\boldsymbol{\psi}(\mathbf{x})$ *such that* $\operatorname{im} \boldsymbol{\psi}$ *is not contained in a hyperplane. Then* $\mathbf{v} \in \ker N^T$. *Conversely, any vector* $\mathbf{v} \in \ker N^T$ *induces a linear conservation law.*

Proof. Let us first assume that $\mathbf{v} \in \ker N^T$. Then

$$\frac{\mathrm{d}}{\mathrm{d}t}\left(\mathbf{v}^T \cdot \mathbf{x}\right) = \mathbf{v}^T N \psi(\mathbf{x}) = \left(N^T \mathbf{v}\right)^T \psi(\mathbf{x}) = 0 \;.$$

If $\mathbf{v}^T \cdot \mathbf{x} = \mathrm{Const}$ is a conservation law, then differentiation with respect to time yields $\left(N^T \mathbf{v}\right)^T \psi(\mathbf{x}) = 0$. Because of our assumption on the function ψ, this implies that $N^T \mathbf{v} = 0$. □ □

By a classical result in linear algebra (the four "fundamental spaces" of a matrix), we have the direct sum decomposition $\mathbb{R}^n = \mathrm{im}\, N \oplus \ker N^T$ which is even an orthogonal decomposition with respect to the standard scalar product. Hence we may consider Lemma 2.2 as a corollary to Lemma 2.4, as the above described invariant manifolds are simply defined by all the linear conservation laws produced by Lemma 2.4.[1]

Remark 2.5. Gatermann and Huber [32] speak of a conservation law only in the case that $v_i \geq 0$ for all components v_i of the vector $\mathbf{v}$. In mathematics, we are not aware of such a restriction and cannot see any physical reasons to impose it.

2.2 Quantifier Elimination and Formula Simplification

Some of our developed methods make fundamental use of Formula simplification and quantifier elimination. In this section, we introduce briefly some of their basic concepts.

A formula of first order logic is a formula that is build from atomic formulas using logical connectives, quantifiers and parentheses. An atomic formula is a formula of the form $P(t_1, t_2, \ldots, t_l)$ where P is a l-ary predicate symbol and $t_1, t_2, \ldots, t_n$ are terms. If a formula does not contain any quantifier, then it is a quantifier free formula. The simplification process for such formula is to find an equivalent formula that is simpler. When formulas are extended by allowing quantification of variables, we may ask for a particular kind of simplification: quantifier elimination. In the 1930's, Tarski proved that any formula containing quantifiers is equivalent to a formula without quantifiers by giving an algorithm to construct a quantifier-free equivalent formula [7]. In this work we use especially the quantifier elimination on real closed field, which is concerned with boolean combination of polynomial equalities and inequalities, where variables are assumed to range over the real numbers.

---

[1] Note that in the special case most relevant for us, namely that each component of ψ is a different monomial, the assumption made in Lemma 2.4 is always satisfied.

There has been during the past decades considerable research on efficient implementation of real quantifier elimination and formula simplification. Some of well-known computational logic tools for solving these problems are listed in the following:

REDLOG[2] [17, 78], which was originally motivated by the efficient implementation of quantifier elimination based on virtual substitution methods [18, 95, 96]. REDLOG also includes cylindrical algebraic decomposition (CAD) and Hermitian quantifier elimination [16, 35, 98] for the reals as well as quantifier elimination for various other domains [77] including the integers [44, 45]. REDLOG is included in the computer algebra system REDUCE, which is an open source.[3] In addition to regular quantifier elimination methods for the reals, REDLOG includes several variants of quantifier elimination [92]. In particular, these variants include *extended quantifier elimination* [97], which additionally yields sample solutions for *existential quantifiers*, and *positive quantifier elimination* [79, 80], which includes powerful simplification techniques based on the knowledge that all considered variables are restricted to positive values. In chemical systems, the region of interest is the positive cone of the state variables and the parameters of interest are known also to be positive. Therefore, positive quantifier elimination is of special importance and will be used for our computations.

QEPCAD [6] is a tool for quantifier elimination that implements partial cylindrical algebraic decomposition. The development of QEPCAD started with the early work of Collins and his collaborators on CAD circa 1973 and continues to this today. QEPCAD is supplemented by another software package called SLFQ for simplifying quantifier-free formulas using CAD. Both QEPCAD[4] and SLFQ[5] are freely available [92].

The SLFQ system uses QEPCAD as a black box for simplifying quantifier-free formulas. QEPCAD is able to simplify formulae, but its time and space requirements become prohibitive when input formulae are large. SLFQ essentially breaks large input formulae into small pieces, uses QEPCAD to simplify the pieces, and starts a process of combining simplified subformulae and applying QEPCAD to simplify the combined subformulae. Eventually this process produces a simplification of the entire initial formula [92].

The well-known commercial computer algebra system Mathematica also includes an efficient CAD-based implementation for real quantifier elimination [92]. The initial developments of the algorithm began circa 2000 by Strzebonski [75, 76]. He developed an CAD-based algorithm for solving systems of strict polynomial inequalities using a simplified projection operator and constructing only rational sample points.

---

[2] http://www.redlog.eu/
[3] http://reduce-algebra.sourceforge.net/
[4] http://www.usna.edu/Users/cs/qepcad/B/QEPCAD.html
[5] http://www.cs.usna.edu/~qepcad/SLFQ/Home.html

Z3 is a new and efficient *satisfiability modulo theories* (SMT) solver that is freely available from Microsoft Research[6]. SMT generalizes boolean satisfiability (SAT) by adding equality reasoning, arithmetic, fixed-size bit-vectors, arrays, quantifiers, and other useful first-order theories. An SMT solver is a tool for deciding the satisfiability (or dually the validity) of formulas in these theories. SMT solvers enable applications such as extended static checking, predicate abstraction, test case generation, and bounded model checking over infinite domains, to mention a few [15]. Z3 uses novel algorithms for quantifier instantiation and theory combination. It is implemented in C++ and integrates a modern DPLL-based SAT solver, a core theory solver that handles equalities and uninterpreted functions, satellite solvers (for arithmetic, arrays, etc.), and an E-matching abstract machine (for quantifiers). For quantifier instantiation Z3 uses a well known approach for quantifier reasoning that works over an E-graph to instantiate quantified variables [15].

RSolver[7] is a new tool for solving quantified inequality constraints. Problems like projecting the solution set of a set of inequality constraints to two dimensions, or the parametric robust stability of linear differential equations can be directly formulated as such constraints. Rsolver is developed by Stefan Ratschan and it is based on reducing the average run-time of algorithms for computationally hard problems by replacing expensive exhaustive search as much as possible by methods for pruning elements from the search space for which it is easy to show that they do not contain solutions. This idea is extended to quantifier inequality constraints for which all (free and bound) variables are bounded to a closed interval [63].

2.3 Linear Programming

Linear programming (LP) is an optimization process that maximize or minimize a linear objective function, subject to linear equality and linear inequality.
We consider the matrix $\mathcal{A}_1 \in \mathbb{R}^{l_1, n}$, the matrix $\mathcal{A}_2 \in \mathbb{R}^{l_2, n}$, the vector $b_1 \in \mathbb{R}^{l_1}$, the vector $b_2 \in \mathbb{R}^{l_2}$, and the vector $c \in \mathbb{R}^n$. A linear program can be expressed in canonical form as shown below:

maximize $c^T x$
subject to

- $\mathcal{A}_1 x = b_1$,

- $\mathcal{A}_2 x \leq b_2$.

---
[6] http://z3.codeplex.com/
[7] http://rsolver.sourceforge.net/

A vector $\hat{x} \in \mathbb{R}^n$ that satisfies each equation $\mathcal{A}_{1i}x = b_{1i}$ and each linear constraint $\mathcal{A}_{2i}x \leq b_{2i}$ is called *feasible solution*. If in addition $c^T\hat{x} \geq c^T x$ for all feasible solutions x, $\hat{x}$ is called then *optimal solution*. The set of all feasible solutions is called the *feasible region*. A linear program is infeasible if its feasible region is empty, otherwise it is called *feasible*. Linear programming problem can be solved using the *simplex method*, which enumerates adjacent vertices of the feasible region such that at each new vertex the objective function improves or unchanged. Another efficient polynomial time algorithm is the *interior point method* [43]. The reader is referred to [68] for more details on theory of linear programming.

There are several free and commercial tools for solving linear programs. In this work we use the commercial software *Gurobi Optimizer*, which is free for academic use. It includes simplex and barrier solvers for linear programming (*LP*) and quadratic programming(*QP*), parallel barrier solver for quadratically constrained programming (*QCP*), as well as parallel mixed-integer linear programming (*MILP*), mixed-integer quadratic programming (*MIQP*) and mixed-integer quadratically constrained programming (*MIQCP*) solvers. The majority of (*LP*) problems can be solved in an efficient way using the *Gurobi*'s state-of-the-art dual simplex algorithm[8].

---

[8]http://www.gurobi.com/

Chapter 3

Generation of Algebraic Data

Chemical reaction systems that follow mass action kinetics can be represented by ordinary differential equations or differential algebraic equations. Their analysis by estimating the values of parameters in these equations is turned out to be difficult for high dimensional systems. However the dynamical behavior of these systems can be analysed using different algebraic methods, which are based on their structures. For applying these methods, the initial task is the generation of algebraic data describing its reaction laws.

3.1 Computation of Basic Algebraic Data

Applying algebraic methods to analyze reaction networks with mass action kinetics requires mainly the generation of three basic algebraic entities describing the reaction laws namely stoichiometric matrix $\mathcal{S}$, kinetic matrix $\mathcal{K}$, and flux vector $v(x, k)$, where x_i's denote the concentrations of chemical species and k_i's represent the reaction rate constants. The stoichiometric matrix describes the occurrence of the species in each reaction, where the rows and columns corresponds to species and reactions respectively. Each entry of the matrix denotes the difference between the production and consumption of molecules of the corresponding species in corresponding reaction. The reversible reaction will be split into two irreversible reactions.

To elucidate its computation from a (bio)-chemical reaction network we consider the Sel´kov model for glycolytic oscillations already discussed in [72]. This simple model involves 5 reactions and 2 species. The first reaction is the autocatalytic production of fructose-1,6-biphosphate consuming adenosine triphosphate (ATP). The other four reactions are the constant inflow and linear outflow of these two species. It is described by the following reaction laws:

Chapter 3. Generation of Algebraic Data

$$2A + B \xrightarrow{k_1} 3A \tag{3.1}$$

$$\underset{k_3}{\overset{k_2}{\rightleftharpoons}} A \tag{3.2}$$

$$\underset{k_5}{\overset{k_4}{\rightleftharpoons}} B \tag{3.3}$$

The stoichiometric matrix of this model is the following 2x5 matrix, where the first row represent the species A and the second row corresponds to B. The columns represent respectively the reactions with rate constants k_1, k_2, k_3, k_4, and k_5.

$$\mathcal{S} = \begin{pmatrix} 1 & 1 & -1 & 0 & 0 \\ -1 & 0 & 0 & 1 & -1 \end{pmatrix}$$

The flux vector $v(x, k)$ (also called *velocity vector*) contains monomials describing the velocity of the reactions. Every monomial is formed by $v_i(x, k_i) = k_i \prod_{j=1}^{m} x^{\alpha_{ij}}$, where m denotes the number of species, and α_{ij}'s denote the stoichiometric coefficients of the reactants in the corresponding reaction i. For the network 3.1 the flux vector is given by:

$$v(x, k) = \begin{pmatrix} k_1 x_1^2 x_2 \\ k_2 \\ k_3 x_1 \\ k_4 \\ k_5 x_2 \end{pmatrix}$$

Another algebraic entity containing information about the velocity of a (bio)-chemical network is the kinetic matrix. It describes this information by encoding the exponents of the flux vector. As the case in stoichiometric matrix the species build the rows and the reaction build the columns. The entries of this matrix present the information whether species is a reactant (entry = stoichiometric coefficient of species) and whether effects consequently the velocity of the reaction or not (entry = 0). The Kinetic matrix of the network 3.1 for instance is:

$$\mathcal{K} = \begin{pmatrix} 2 & 0 & 1 & 0 & 0 \\ 1 & 0 & 0 & 0 & 1 \end{pmatrix}$$

3.2 Flux Cone and Extreme Currents

To analyze a (bio)-chemical system one is interested in the stationary reaction behavior, which is observable in experiments, i.e one investigates the solution set of

$$\mathcal{S}v(x,k) = 0. \tag{3.4}$$

where $\mathcal{S}$ represent the stoichiometric matrix and $v(x,k)$ the flux vector. As long as we split each reversible reaction into two irreversible reactions (forward and backward directions) the flux through these reactions must be greater or equal to zero, i.e

$$v(x,k) \geq 0 \tag{3.5}$$

The set of all possible stationary solutions over a (bio)-chemical network $\mathcal{N}$ that fulfill the equation (3.4) and the constraint (3.5) defines the convex polyhedral cone *flux cone* [12, 43]. The minimal set of generating vectors $\mathcal{E}$, which can geometrically be interpreted as the edges of the flux cone are known in chemistry as *extreme fluxes* or *extreme currents*. The Fig. 3.1 depicts the extreme currents of a flux cone[1].

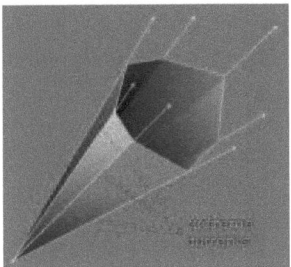

FIGURE 3.1: Extreme currents of flux cone

Each flux vector satisfying the steady state equations can be represented in the flux space as linear combination of the extreme currents $\mathcal{E}$ with nonnegative coefficients j_i called *convex parameters*. In reaction networks, as the reversible reactions are split into two, so for each split there appears a spurious cycle [91] in the set of extreme currents, which is an extreme current denoting only the forward and backward reaction components of the reversible reaction which can be removed as an additional step. The reversibility of reactions in the network can be tackled by different ways. In extreme currents mentioned above the reversible reactions are split but in elementary flux modes [69] they are not split and in extreme pathways [67] some may be split. The different ways to split the

---
[1]From http://www.csb.ethz.ch/research/structural (edited figure)

reactions affects the construction of stoichiometric matrix and hence different methods describe the cone in different vector spaces due to presence of reversible reactions [48, 91]. It also affects the construction of cone, for extreme currents computation the cone is always pointed whereas in elementary flux modes and extreme pathways it may neither be pointed nor remain in non-negative orthant [48]. This coordinate transformation of network to cone can also be used for stability analysis [30].

The flux cone of the Sel´kov model (Network 3.1) is spanned by the three extreme currents

$$\mathcal{E}_1 = \begin{pmatrix} 0 & 1 & 1 & 0 & 0 \end{pmatrix},$$
$$\mathcal{E}_2 = \begin{pmatrix} 0 & 0 & 0 & 1 & 1 \end{pmatrix},$$
$$\mathcal{E}_3 = \begin{pmatrix} 1 & 0 & 1 & 1 & 0 \end{pmatrix}.$$

3.3 Computation of Jacobian Matrix Using Convex Coordinates

The basic idea to transform the Jacobian coordinates from concentration space to reaction space is described by Gatermann et al. in [30]. As the temporal behavior of chemical species during a reaction sequence with mass action kinetics can also be given by the Equation 3.6, where φ contains the pure monomials of $v(x,k)$ without the parameters k_i

$$\dot{x} = \mathcal{S}v(x,k) = \mathcal{S}diag(k)\varphi, \qquad (3.6)$$

and by defining some rescalings they proved that the new Jacobian matrix is the product of the Jacobian of the flux vector in reaction space $\mathcal{Z}$ and the diagonal matrix of the inverse of the concentrations.

$$\text{Jac}(x) = \widehat{\text{Jac}}(\mathcal{Z})\text{diag}(1/x_1, ..., 1/x_m), \qquad (3.7)$$

with $\widehat{\text{Jac}}(\mathcal{Z}) = \mathcal{S}\text{diag}(\mathcal{Z})\mathcal{K}^t$.

If x is a steady state we transform into the convex coordinates j_i with $\mathcal{Z} = \sum_i j_i \mathcal{E}_i$. Finally this yields the Jacobian

$$\text{Jac}(x) = \widehat{\text{Jac}}(j)\text{diag}(1/x_1, ..., 1/x_m), \qquad (3.8)$$

Chapter 3. *Generation of Algebraic Data*

with $\widehat{\text{Jac}}(j) = \mathcal{S}\text{diag}(\sum_i j_i \mathcal{E}_i)\mathcal{K}^t$.

Since the most information on stability is included in the first three matrix and the sign pattern of Jac(x) is the same as the sign pattern of $\widehat{\text{Jac}}(j)$ [72], it is convenient to decompose the Jacobian matrix and to consider only the matrix $\widehat{\text{Jac}}(j)$ which is depend only on the convex parameters. The Jacobian in convex coordinates of the Sel´kov model is:

$$\widehat{\text{Jac}}(j) = \begin{pmatrix} j_3 - j_1 & j_3 \\ -2j_3 & -j_2 - j_3 \end{pmatrix},$$

and

$$\text{Jac}(x) = \begin{pmatrix} \frac{j_3 - j_1}{x_1} & \frac{j_3}{x_2} \\ \frac{-2j_3}{x_1} & \frac{-j_2 - j_3}{x_2} \end{pmatrix}.$$

3.4 Algebraic Data for Graph-Theoretic Representation of the Reaction Systems

Based on the notion that reaction networks can be modelled using differential equations, the rate of change of every metabolite $\dot{x}$ in such a network can be represented as

$$\dot{x} = Y I_a I_k \Psi(x), \tag{3.9}$$

where $I_k \Psi(x)$ denotes the vector of monomials describing the flux of the reactions and $Y I_a$ denotes the stoichiometric matrix. Let the network has l reactions and m species. Here we follow the notation and terminology used in [29–31]. The above differential equation can be represented by the use of two graphs, a weighted directed graph and a bipartite undirected graph. In addition to this, both sides of a reaction (i.e. products and reactants) are arranged in form of complexes (complex may consists of a single species or combination of species). Let the network has n complexes. In the directed graph, there exists directed edge between the two complexes describing a reaction. The edge weight is the rate constant of the reaction. From this graph two incidence matrices are defined I_a and I_k. The I_a is a n-by-l matrix and has the information whether the complex is present as a reactant (entry -1) or product (entry 1) vertex of the graph. The I_k is a l-by-n matrix that has non zero entries only for initial vertices where the entry is the weight of the edge which is the rate constant of the reaction. Each complex can be mapped with a monomial where its exponent is its stoichiometry and $\Psi(x)$ provides this mapping information. The following matrices represent the I_a matrix and I_k of the Sel´kov model (Network 3.1).

$$I_a = \begin{pmatrix} -1 & 0 & 0 & 0 & 0 \\ 1 & 0 & 0 & 0 & 0 \\ 0 & -1 & 1 & -1 & 1 \\ 0 & 1 & -1 & 0 & 0 \\ 0 & 0 & 0 & 1 & -1 \end{pmatrix}$$

$$I_k = \begin{pmatrix} k_1 & 0 & 0 & 0 & 0 \\ 0 & 0 & k_2 & 0 & 0 \\ 0 & 0 & 0 & k_3 & 0 \\ 0 & 0 & k_4 & 0 & 0 \\ 0 & 0 & 0 & 0 & k_5 \end{pmatrix}$$

The bipartite graph contains the set of complexes and the set of chemical species as vertices. If the complex contains a species then there exists an edge from that complex to the chemical species with the edge weight equals to the stoichiometry of the chemical species. The adjacency matrix of this graph is denoted by Y matrix (m-by-n) (in Equation 3.9). The Y matrix of the Network 3.1 is:

$$Y = \begin{pmatrix} 2 & 3 & 0 & 1 & 0 \\ 1 & 0 & 0 & 0 & 1 \end{pmatrix}$$

Additionally, the number of linkage class which can be utilized for example for the computation of deficiency, can be found from the weighted directed graph i.e. it is the set of connected complexes. Further details concerning complexes and linkage classes can be found in [25]. The benefit of this graph-theoretic approach leads to the solutions having a graph theoretic meaning [31] and stability analysis [30].

3.5 Deficiency Value of the Reaction Network

Deficiency is a non negative integer for a reaction network which is an invariant of the network. In this context, two well known theorems are available which are Deficiency

Chapter 3. *Generation of Algebraic Data* 19

Zero and Deficiency One theorems respectively [25]. The first step in this direction is the computation of this deficiency value which is given by following formula:

$$\delta = n - t - s \tag{3.10}$$

where n is the number of complexes in the network, t is number of linkage classes, s is the rank of network. In addition, to this the deficiency can also be computed using the formula from the graph theoretic representation [29]:

$$\delta = Rank(I_a) - Rank(YI_a) \tag{3.11}$$

This deficiency value enables to classify the reaction networks into the kind of dynamics they can possibly exhibit.

3.6 *PoCaB*: A Software Infrastructure to Explore Algebraic Methods for (Bio)-Chemical Reaction Networks

In this section we describe a general framework called *PoCaB* to generate relevant algebraic entities out of the (bio)-chemical network description such as stoichiometric matrices and their factorizations, kinetic matrices, extreme currents, polynomial systems, deficiencies and differential equations. We also use *PoCaB* to extract and compute algebraic entities form different biological models obtained from two publicly available databases and provide the results as large *derived database*[2] of examples that can be used by people working in computer algebra to benchmark their algorithms. The framework can serve as a manual for chemists and biologists to apply diverse algebraic methods in a systematic manner and interpret the results. This will also help us to formulate subsequent computational questions on the applicability, pros and cons of such methods to analyse the large and diverse datasets.

3.6.1 Representation of Reaction Networks

To enable the computational analysis of a chemical networks the reactions should be presented in a format that enables its accurate representation and allows the computational extraction of needed data. For our computations we use the XML based and in biological research widely used format SBML (System Biology Markup Language) to communicate biochemical reaction network consisting of metabolic pathways, signaling

---
[2]available at http://pocab.cg.cs.uni-bonn.de/

Chapter 3. *Generation of Algebraic Data* 20

```
<?xml version="1.0" encoding="UTF-8"?>
- <sbml xmlns="http://www.sbml.org/sbml/level2/version4" metaid="_042589" level="2" version="4">
 - <model metaid="_000001" id="Goldbeter1991_MinMitOscil_ExplInact" name="Goldbeter1991_MinMitOscil_ExplInact">
    + <notes>
    + <annotation>
    + <listOfCompartments>
    + <listOfSpecies>
    + <listOfParameters>
    + <listOfRules>
    + <listOfReactions>
    </model>
 </sbml>
```

FIGURE 3.2: Components of SBML file

pathways, gene regulation pathways, etc and is also software independent (Fig. 3.2). Further details about the various specifications can be found in the SBML tutorial [40].

3.6.2 Database of Algebraic Entities

3.6.2.1 Data Source

Biomodels: We selected 270 biochemical reaction networks from Biomodels database in SBML. These models can be browsed by name of the disease, biological process and molecular complex.

KEGG: The KEGG database is another repository of biological pathways. The KEGG pathways can be downloaded in KGML format. For our analysis we downloaded a precompiled list of KEGG files in SBML format[3]. We selected 103 models with organism code hsa (*Homo sapiens*). In addition, if downloaded in KGML format the files can be converted to SBML using KEGG translator [99].

However, our framework is not restricted to these databases but can be used for all sources that provide data in SBML form.

3.6.2.2 Software Workflow and Components

Pre-processing Step: From the datasource (as described in Sect. 3.6.2.1) the networks were downloaded. Although the models in these databases are annotated and curated, still as a part of general framework we have a possibility to balance the reactions. It removes stoichiometric inconsistencies present in the model [34] and works only when the annotation of model is correct and the chemical formula of species can be found. This works on the principle of mixed integer linear programming (MILP) [34]. This is done automatically using the Subliminal toolbox [82] which implements the

---

[3]downloaded from http://www.systems-biology.org/resources/model-repositories/000275.html

above steps. We present the results for this only for Biomodels database. However, this step is optional and we report results with and without balancing.

Main Steps:

1. The files are parsed by a Java based program to generate the Y, I_a, stoichiometric matrix (YI_a) and kinetic order matrix along with the basic information about the model concerning the number of species, reactions, complexes, rank of stoichiometric matrix and nullity of stoichiometric matrix respectively. While computing the various matrices the reversible reactions are split into forward and backward reaction, it increases the dimension of the stoichiometric matrix by one for every reversible reaction. To parse the SBML program using Java, the JSBML library is used [21]. The graph theoretic representation of the network (cf. Sect. 3.4) was done using JGraphT Java Library[4].

2. The deficiency of the networks were computed using the *ERNEST* library [74], which is a Matlab based program. This program also splits the reversible reactions into two separate reactions, so the deficiency value can be directly computed from the SBML file. In addition to this the tool also tests whether the Deficiency zero or one theorems are applicable and presents the results. We report only the deficiency value and for additional conditions for deficiency theorems, the *ERNEST* can be used e.g. weak reversibility, linkage classes. The deficiency can also be calculated from Equation 3.11 and is found to be same as *ERNEST* for all our files.

3. The stoichiometric matrix acts as an input to the tools allowing the polyhedral computations e.g. *polco* and *polymake*. the Java based tool POLCO[5] implements the double description algorithm [84] to compute the extreme currents. *polymake*[6] is a computer algebra tool that was written in Perl and C++ and designed for the algorithmic treatment of polytopes and polyhedra [33]. One of the advantages of using this tool is the different choice of algorithms for convex hull computation. Further theoretical exploration into different properties of polyhedrons can be found in [101]. The output is a matrix (*o*-by-*l*) with *o* denoting the number of extreme currents.

4. The reaction network, can be represented by inequalities and equations. One benefit of this representation is that different constraints can be put on individual reactions. In biology, the networks operate under different constraints[56], one

---
[4]http://www.jgrapht.org
[5]http://www.csb.ethz.ch/tools/polco
[6]http://www.polymake.org/doku.php

Chapter 3. *Generation of Algebraic Data*

important constraint is the effect of gene regulation in which some genes are differentially expressed [14]. The changes in gene expression levels affect the reaction rates as the reactions are governed by enzymes which are gene products. One of the ways to model this phenomenon is to use inequalities [87]. The following formalism illustrates the above points:

$$YI_a I_k \Psi(x) = 0 \qquad (3.12)$$

$$I_k \Psi(x) \geq 0 \qquad (3.13)$$

$$I_k \Psi(x) \leq \beta_i \quad (i = 0, 1, \ldots, l-1) \qquad (3.14)$$

where l denotes the number of reactions. Equation 3.12) denotes the steady state, while Equation (3.14) denotes the constraint on the flux of a reaction and hence is optional. In the current analysis we have not accounted for any constraints, so in a way the flux cone computed is somewhat maximal where all reactions occur at their maximal rate. We systematically generated this type of file for all the examples (also a part of our derived database) so that it can be communicated to other computer algebra programs e.g. polymake. This file is also directly imported into polymake.

3.6.2.3 Content of Database

1. The generated Y, I_a, I_K, and $I_K\Psi(x)$, stoichiometric matrices (YI_a), kinetic order matrix, extreme currents along with the polynomial system and the differential equation files are stored as text files using delimiter (,) to separate the elements of the matrices.

2. In the above files mainly the I_K, $I_K\Psi(x)$, polynomial system and differential equation files, the species name are mapped to certain variable. The rate constants also mapped to corresponding reactions where they occur. This mapping information is present as a Mapping file. This file also contains the information about the reactions and the species involved in the network.

3.6.2.4 Statistical Summary

We described a database having algebraic entities derived from biological reaction networks. One important feature of this database is that it is extensible, for instance, the examples can be annotated with stability information, steady states, oscillations, etc. The matrices, flux cone and differential equations which were computed provide the basic framework for such kind of advanced analysis. From the results it can be seen that the number of extreme currents does not always correspond to the size of the network as seen in Biomodels database. Also the size of network doesn't relate to the deficiency also pointed out in [25]. But there are also some models displaying high deficiency with large dimensions and there is a need of improved algorithms or new approaches to address such systems. It can be seen around 71.1% (unbalanced) and 77.1 % (balanced) models in Biomodels correspond to deficiency one or zero. Similarly 88 % (unbalanced) of KEGG models correspond to this criteria, this implies the existing deficiency theorems are applicable to a large extent. The effect of balancing the reactions can also be seen in Table 3.1, there may be changes in the number of species and reactions during the balancing and this affects the number of extreme currents and deficiency computation. As our derived database contain diverse examples, it provides a corpus to test and benchmark different algebraic methods and designate the methods working for a particular class of examples. This will eventually lead to partitioning the database into classes which may be suitable for some methods and unsuitable for others. A natural partitioning occurs for examples with deficiency zero or one and there exists theorems to apply on such examples. Additional type of partitioning can be based on the dimension of various matrices, number of extreme currents and one such possibility is presented in Table 3.1 which is based on the number of reactions upto 10 and 50. The identifier used in the source database is same as that of our database (name of the file), which will aid in linking biological information along with the networks, so the theoretical results can be corroborated with real biological phenomena. As the steps described to create such a database are relatively simple so, it will enable chemists/biologists to use it to develop some hypotheses about their future experiments. Furthermore, to simplify this there exists a tool called CellDesigner [26] with graphical interface to encode reaction networks and export them to SBML format. From a biological perspective we intend to understand biological function from a set of reactions but there is no consensus about the boundary of system under consideration and it varies widely among the researchers. So, for the same function there may exist several competing models and the goal should be then to discriminate models by finding a model/set of models which are more close to reality. This exercise of model discrimination can be also performed with our database as the source database specifically the Biomodels contain related models for a biological property and it will be desirable to see in future if the algebraic techniques can distinguish them from others.

TABLE 3.1: Summary of results in Biomodels and KEGG database

	UnBalanced Biomodels	KEGG	Balanced Biomodels
Number of Models	270	103	236
Maximum number of reactions* in a model	194	132	194
Maximum number of species in a model	120	139	120
Number of models with reactions upto 10	94	48	117
Number of models with reactions upto 50	237	97	213
Maximum number of EC* in a model	5130	282	5130
Dimension of SM* with maximum EC	17×48	139×132	19×48
Models with deficiency = 0	159 (58.8%)	80 (77.6%)	154 (65.2%)
Models with deficiency = 1	33 (12.2%)	11 (10.6%)	28 (11.8%)
Highest Deficiency	63	24	63
SM with highest deficiency	36×94	139×132	39×94
Maximum rank of SM	94	67	94
Maximum nullity of SM	100	65	100
Number of models with zero EC	24 (8.88%)	38 (36.8%)	50 (21.1%)

*The reactions here refer to columns of SM. SM = Stoichiometric matrix.

EC=Extreme Currents.

Chapter 4

Detection of Hopf Bifurcations Using Convex Coordinates

In this chapter, we present efficient algorithmic methods to detect Hopf bifurcations in (bio)-chemical reaction systems with linear constraints. Hence they yield information on local oscillatory behavior of these systems. Our methods use the representation of the systems on convex coordinates arising from stoichiometric network analysis. One of our method reduces the problem of determining the existence of Hopf bifurcation fixed points to a first-order formula over the ordered field of the reals that then can be decided using packages from computational logic. The second method uses ideas from tropical geometry to formulate more efficient approach that is incomplete in theory but worked very well for the complex models that we have attempted.

4.1 Hopf Bifurcations and Invariant Manifolds

4.1.1 Conditions for Existence of Hopf Bifurcations

Consider a parameterized autonomous ordinary differential equation of the form $\dot{x} = f(u, x)$ with a scalar parameter u. By a classical result of Hopf, at the point (u_0, x_0), this system exhibits a Hopf bifurcation, i.e. an equilibrium transforms into a limit cycle, if $f(u_0, x_0) = 0$ and if the Jacobian $D_x f(u_0, x_0)$ has a simple pair of purely imaginary eigenvalues and no other eigenvalues with zero real parts [37, Thm. 3.4.2].[1] The proof of this result is based on the center manifold theorem. From a physical point of view, the

---
[1] We ignore here the non-degeneracy condition that this pair of eigenvalues crosses the imaginary axis transversally, as it is always satisfied in realistic models.

Chapter 4. Detection of Hopf Bifurcation Using Convex Coordinates

most interesting case is that the unstable manifold of the equilibrium (u_0, x_0) is empty. However, for the mere existence of a Hopf bifurcation, this assumption is not necessary.

In [23], it is shown that for a parameterized vector field $f(u, x)$ and the autonomous ordinary differential system associated with it, there is a semi-algebraic description of the set of parameter values for which a Hopf bifurcation (with an empty unstable manifold) occurs. Specifically, this semi-algebraic description can be expressed by the following first-order formula:

$$\exists x (f_1(u,x) = 0 \land f_2(u,x) = 0 \land \cdots \land f_n(u,x) = 0$$
$$\land \, a_n > 0 \land \Delta_{n-1}(u,x) = 0 \land \Delta_{n-2}(u,x) > 0 \land \cdots \land \Delta_1(u,x) > 0) \quad (4.1)$$

In this formula a_n is $(-1)^n$ times the Jacobian determinant of the matrix $Df(u,x)$, and $\Delta_i(u,x)$ is the ith Hurwitz determinant of the characteristic polynomial of the same matrix $Df(u,x)$.

The proof uses a formula of Orlando [54], which is also discussed in several monographs, e.g. in [27] and [59]. However, a closer inspection of the two parts of the proof of [23, Theorem 3.5] shows the following: for a fixed point (given in possibly parameterized form) the condition that there is a pair of purely imaginary eigenvalues is given by the condition $\Delta_{n-1}(u,x) = 0$ and the condition that each other eigenvalue has a negative real part is given by $\Delta_{n-2}(u,x) > 0 \land \cdots \land \Delta_1(u,x) > 0$. This statement (without referring to parameters explicitly) is also contained in [100, Theorem 2], in which a different proof technique is used.

Therefore, if we drop the condition for Hopf bifurcation points that they have empty unstable manifolds, a semi-algebraic description of the set of parameter values for which a Hopf bifurcation occurs for the system is given by the following formula:

$$\exists x (f_1(u,x) = 0 \land f_2(u,x) = 0 \land \cdots \land f_n(u,x) = 0$$
$$\land \, a_n > 0 \land \Delta_{n-1}(u,x) = 0) \quad (4.2)$$

Notice that when the quantifier elimination procedure yields sample points for existentially quantified formulae—as is the case for the virtual-substitution based method provided by REDLOG—then the condition $\Delta_{n-2}(u,x) > 0 \land \cdots \land \Delta_1(u,x) > 0$ can be tested for the sample points later on, i.e. one can then test whether this Hopf bifurcation fixed point has an empty unstable manifold.

Chapter 4. Detection of Hopf Bifurcation Using Convex Coordinates

Example: Lorenz system The famous "Lorenz system" [37, 49, 62] is given by the following system of ODEs:

$$\frac{d}{dt}x(t) = \alpha\left(y(t) - x(t)\right) \quad (4.3)$$

$$\frac{d}{dt}y(t) = r\,x(t) - y(t) - x(t)\,z(t) \quad (4.4)$$

$$\frac{d}{dt}z(t) = x(t)\,y(t) - \beta\,z(t) \quad (4.5)$$

It is named after Edward Lorenz at MIT, who first investigated this system as a simple model arising in connection with fluid convection.

After imposing positivity conditions on the parameters the following answer is obtained using a combination of REDLOG and formula simplification using SLFQ for the test of a Hopf bifurcation fixed point:

$$(-\alpha^2 - \alpha\beta + \alpha r - 3\alpha - \beta r - r = 0 \vee -\alpha\beta + \alpha r - \alpha - \beta^2 - \beta = 0) \wedge$$
$$-\alpha^2 - \alpha\beta + \alpha r - 3\alpha - \beta r - r \leq 0 \wedge$$
$$\beta > 0 \wedge \alpha > 0 \wedge -\alpha\beta + \alpha r - \alpha - \beta^2 - \beta \geq 0 \quad (4.6)$$

When testing for Hopf bifurcation fixed points with empty unstable manifolds, we obtain the following formulae:

$$\alpha^2 + \alpha\beta - \alpha r + 3\alpha + \beta r + r = 0 \wedge$$
$$\alpha r - \alpha - \beta^2 - \beta \geq 0 \wedge$$
$$2\alpha - 1 \geq 0 \wedge \beta > 0 \quad (4.7)$$

These two formulae are not equivalent, and therefore, for the case of the Lorenz system not all Hopf bifurcation fixed points have unstable empty manifolds.

4.1.2 Reduction to Invariant Manifolds

As already discussed in Sect. 2.1.2, chemical reaction systems with linear conservation laws can easily be generalized to pseudolinear ordinary differential equations. However the existence of these constraints makes the Jacobian matrices singular and thus leads to incorrect computations of Hopf bifurcations. We present here a method to tackle these singularities by reduction to invariant manifolds. The following material represents a slight generalization of results already well-known for systems in reaction kinetics (see, e.g. [32] and references therein).

If a dynamical system admits invariant manifolds, we may consider a system of lower dimension by reducing to such a manifold. However, in general it may not be possible to explicitly derive the reduced system. Nevertheless, for many purposes, such as stability or bifurcation analysis, one can easily reduce to smaller matrices. The following result describes such a reduction process in the linear case. It represents an elementary exercise in basic linear algebra. To avoid the inversion of matrices, we consider $\mathbb{R}^n$ here to be a Euclidean space with respect to the standard scalar product.

Lemma 4.1. *Let A be the matrix of a linear mapping $\mathbb{R}^n \to \mathbb{R}^n$ for the standard basis, and let $\mathcal{U} \subseteq \mathbb{R}^n$ be a k-dimensional A-invariant subspace. If the columns of the matrix $W \in \mathbb{R}^{n \times k}$ define an orthonormal basis of $\mathcal{U}$, then the restriction of the mapping to the subspace $\mathcal{U}$ with respect to the basis defined by W is given by the matrix $W^T A W \in \mathbb{R}^{k \times k}$.*

Proof. Considered as a linear map $\mathbb{R}^k \to \mathcal{U} \subseteq \mathbb{R}^n$, the matrix W defines a parametrization of $\mathcal{U}$ with inverse $W^T : \mathcal{U} \to \mathbb{R}^k$. Indeed, $W^T W = \mathbb{1}_k$, since the columns of W are orthonormal. If $\mathbf{v} \in \mathcal{U}$, then $\mathbf{v} = W\mathbf{w}$ for some vector $\mathbf{w} \in \mathbb{R}^k$ and thus $W^T \mathbf{v} = (W^T W)\mathbf{w} = \mathbf{w}$ implying that $(WW^T)\mathbf{v} = W\mathbf{w} = \mathbf{v}$, i.e. the matrix $WW^T \in \mathbb{R}^{n \times n}$ describes $\mathrm{id}_{\mathcal{U}}$. By standard linear algebra, the matrix $W^T A W$ therefore describes the restriction of A to $\mathcal{U}$. □ □

As a simple application, we note that in the case of a pseudolinear system $\dot{\mathbf{x}} = N\psi(\mathbf{x})$ the stability properties of an equilibrium $\mathbf{x}_e$ of the pseudolinear system $\dot{\mathbf{x}} = N\psi(\mathbf{x})$ are determined by the eigenstructure of the reduced Jacobian

$$J = W^T N \mathrm{Jac}\big(\psi(\mathbf{x}_e)\big) W \in \mathbb{R}^{k \times k}$$

where the columns of W form an orthonormal basis of $\mathrm{im}\, N$. If parameters are present, then for a bifurcation analysis the eigenstructure of this matrix and not of the full Jacobian (which is an n-dimensional matrix), is relevant.

4.1.3 Stability and Bifurcations for Semi-Explicit DAEs

The considerations indicated in the previous section can be easily extended to more general situations, as they appear in the theory of DAEs. For simplicity (and because

Chapter 4. Detection of Hopf Bifurcation Using Convex Coordinates

it suffices for our purposes), we assume that we are dealing with an autonomous system in the semi-explicit form

$$\dot{\mathbf{x}} = \mathbf{f}(\mathbf{x}), \qquad 0 = \mathbf{g}(\mathbf{x}) \qquad (4.8)$$

where $\mathbf{f} : \mathbb{R}^n \to \mathbb{R}^n$ and $\mathbf{g} : \mathbb{R}^n \to \mathbb{R}^{n-k}$. Furthermore, we assume that the above system of ordinary differential equations is involutive,[2] i.e. that it already contains all its integrability conditions. This assumption is equivalent to the existence of a matrix valued function $M(\mathbf{x})$ such that

$$\mathrm{Jac}\bigl(\mathbf{g}(\mathbf{x})\bigr) \cdot \mathbf{f}(\mathbf{x}) = M(\mathbf{x}) \cdot \mathbf{g}(\mathbf{x}). \qquad (4.9)$$

Therefore, one may say that the components of $\mathbf{g}$ are *weak* conservation laws, as their time derivatives vanish modulo the constraint equations $\mathbf{g}(\mathbf{x}) = 0$.

Let $\mathbf{x}_e$ be an equilibrium of (4.8), i.e. we have $\mathbf{f}(\mathbf{x}_e) = 0$ and $\mathbf{g}(\mathbf{x}_e) = 0$. We introduce the real matrices

$$A = \mathrm{Jac}\bigl(\mathbf{f}(\mathbf{x}_e)\bigr) \in \mathbb{R}^{n \times n}, \quad B = \mathrm{Jac}\bigl(\mathbf{g}(\mathbf{x}_e)\bigr) \in \mathbb{R}^{(n-k) \times n}.$$

For simplicity, we assume in the following that the matrix B has full rank (or, in other words, that our algebraic constraints are independent) and thus that $\ker B$ is a k-dimensional subspace. The proof of the next result clearly demonstrates why the assumption that the system (4.8) is involutive is important, as the relation (4.9) is crucial for it.

Lemma 4.2. *The subspace $\ker B$ is A-invariant.*

Proof. Set $\bar{M} = M(\mathbf{x}_e)$. Differentiating (4.9) and evaluating the result at $\mathbf{x} = \mathbf{x}_e$ yields the relation $BA = \bar{M}B$. Thus, if $\mathbf{v} \in \ker B$, then also $A\mathbf{v} \in \ker B$ because $B(A\mathbf{v}) = \bar{M}(B\mathbf{v}) = 0$. □ □

In the case that (4.8) is a linear system, i.e. we may write $\mathbf{f}(\mathbf{x}) = A\mathbf{x}$ and $\mathbf{g}(\mathbf{x}) = B\mathbf{x}$ by assuming that $\mathbf{x}_e = 0$, we can easily revert the argument in the proof of Lemma 4.2 and thus conclude that now (4.8) is involutive, if and only if $\ker B$ is A-invariant.

Proposition 4.3. *Let the columns of the matrix $W \in \mathbb{R}^{n \times k}$ define an orthonormal basis of $\ker B$. The linear stability of the equilibrium $\mathbf{x}_e$ is then decided by the eigenstructure of the matrix $W^T A W$.*

Proof. Linearization around the equilibrium $\mathbf{x}_e$ yields the associated variational system $\dot{\mathbf{z}} = A\mathbf{z}, B\mathbf{z} = 0$. We complete W to an orthogonal matrix $\widehat{W}$ by adding some further

---
[2]See [70] for an introduction to the theory of involutive systems.

columns and perform the coordinate transformation $\mathbf{z} = \widehat{W}\mathbf{y}$. This yields the system $\dot{\mathbf{y}} = \widehat{W}^T A \widehat{W} \mathbf{y}$, $B\widehat{W}\mathbf{y} = 0$. Because the columns of W span $\ker B$ by construction, the second equation implies that only the upper k components of $\mathbf{y}$ may be different from zero. Furthermore, Lemma 4.2 implies that the matrix $\widehat{W}^T A \widehat{W} \mathbf{y}$ is in block triangular form with the left upper $k \times k$ block given by $W^T AW$. If we denote the upper part of $\mathbf{y}$ by $\tilde{\mathbf{y}}$, we thereby obtain the equivalent reduced system $\dot{\tilde{\mathbf{y}}} = W^T AW \tilde{\mathbf{y}}$ which implies our claim. $\square$ $\square$

Let $\mathbf{v} \in \mathbb{R}^k$ be a (generalized) eigenvector of the reduced matrix $W^T AW$, i.e. we have $(W^T AW - \lambda \mathbb{1}_k)^\ell \mathbf{v} = 0$ for some $\ell > 0$ and $\lambda \in \mathbb{R}$. Because $W^T W = \mathbb{1}_k$ and WW^T defines the identity map on $\ker B$ (see the proof of Lemma 4.1), we obtain $W^T (A - \lambda \mathbb{1}_n)^\ell W \mathbf{v} = 0$ implying that $W\mathbf{v} \in \mathbb{R}^n$ is a (generalized) eigenvector of A for the same eigenvalue λ, since the matrix W^T defines an injective map. Therefore every eigenvalue of the reduced matrix $W^T AW$ is also an eigenvalue of A.

It is also not difficult to interpret the remaining (generalized) eigenvectors of A. By construction, they are transversal to the constraint manifold defined by $\mathbf{g}(\mathbf{x}) = 0$ and they describe whether this manifold is attractive or repulsive for the flow of the unconstrained system $\dot{\mathbf{x}} = \mathbf{f}(\mathbf{x})$. While this is for example of considerable importance to the numerical integration of (4.8), as it describes the drift off the constraint manifold arising from rounding and discretization errors, it has no influence on the stability of the exact flow of (4.8).

The irrelevance of the remaining (generalized) eigenvectors of A also becomes apparent from the following argument. Recall that the differential part of (4.8) defines what is often called an *underlying differential equation* for the DAE, i.e. an unconstrained differential equation which possesses for initial data satisfying the constraints the same solution as the DAE. Consider now the modified system obtained by adding to the right hand side of the differential part an arbitrary linear combination of the algebraic part. It is easy to see that the arising DAE (which simply has a different underlying equation)

$$\dot{\mathbf{x}} = \mathbf{f}(\mathbf{x}) + L(\mathbf{x})\mathbf{g}(\mathbf{x}), \qquad 0 = \mathbf{g}(\mathbf{x}),$$

where $L(\mathbf{x})$ is a matrix valued function of appropriate dimensions, possesses exactly the same solutions as (4.8); in particular $\mathbf{x}_e$ is still an equilibrium. If we proceed as above with the linear stability analysis of $\mathbf{x}_e$, the matrix B remains unchanged, whereas A is transformed into the modified matrix $\tilde{A} = A + \bar{L}B$ with $\bar{L} = L(\mathbf{x}_e)$. Obviously, $\ker B$ is also $\tilde{A}$-invariant, and furthermore $W^T \tilde{A} W = W^T AW$, if the columns of W form a basis of $\ker B$ as in Proposition 4.3.

Therefore, all (generalised) eigenvectors lying in ker B are equal for A and $\tilde{A}$, so the stability of $\mathbf{x}_e$ is not affected by this transformation. However, the remaining (generalised) eigenvectors may change arbitrarily. One can for example show that by a suitable choice of the matrix L one may always achieve that the constraint manifold becomes attractive.

4.2 $HoCoQ$: An Algorithm for Computing <u>H</u>opf Bifurcations using <u>C</u>onvex Coordinates and <u>Q</u>uantifier Elimination

In this section, we present an algorithmic approach for computing the Hopf bifurcations in chemical systems using convex coordinates instead of concentration coordinates. It is based on two methods already presented in this book: stoichiometric network analysis and manifold reduction for systems with conservation laws. It also makes fundamental use of real quantifier elimination on a real closed field. Figure 4.1 elucidates the workflow of the algorithm, which is explained in detail in the following subsections and in the pseudo-code presented in Algo. 4.2.5.

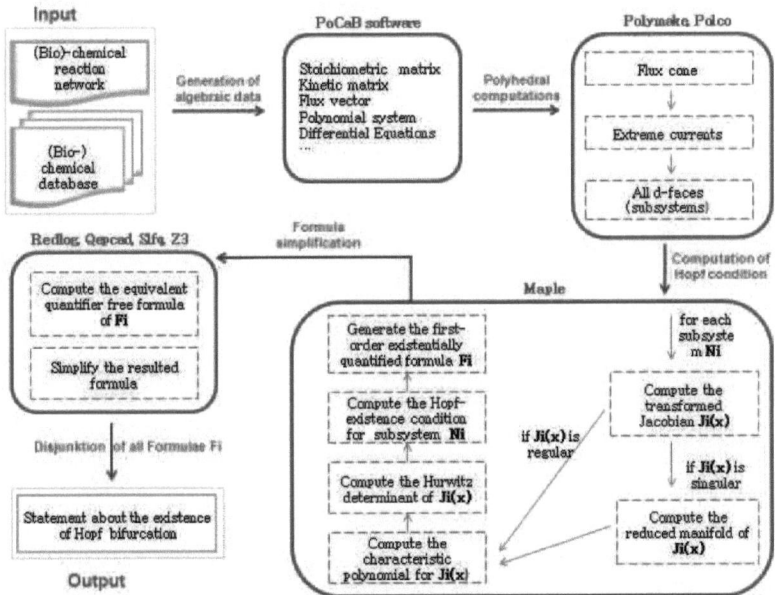

FIGURE 4.1: $HoCoQ$ method

4.2.1 Pre-processing

To begin the analysis of a chemical network in convex coordinates we need two significant pieces of information namely stoichiometric matrix $\mathcal{S}$ and the kinetic $\mathcal{K}$. As already mentioned in Sect. 3.6 we use the SBML-format for the presentation of the chemical reaction networks and as a pre-processing step we use our software $PoCaB$ to generate the necessary algebraic data.

4.2.2 Polyhedral Computations

The advantage of stoichiometric network analysis is the ability to analyze subnetworks separately instead of analyzing the whole complex network. The first step in the analysis is the computation of extreme currents. We must therefore include algorithms that are capable of dealing with polyhedral computations. There are several software packages for such computations and in computational geometry in particular. We use in our current implementation two efficient tools, namely POLCO and POLYMAKE (see Sect. 3.6.2.2).

Enumerating extreme currents $\mathcal{E}$ is the basis for simplifying the analysis of chemical networks by decomposing the network into minimal steady-state generating subnetworks. The influence of a subnetwork on the full network dynamics (i.e., how much the given subnetwork plays a part in creating a certain steady state) depends on the convex parameters j_i [12, 20]. From a chemical perspective, Hopf bifurcations occur mostly in the spaces formed by two or three adjacent extreme currents, i.e detecting Hopf bifurcations in subsystems can be restricted to the subsystems that are formed by combining 2-faces or 3-faces of the flux cone. As step 3 of our algorithms, we compute all subsystems generated by the 2- and 3-faces using POLYMAKE. Our algorithm can also handle d-faces for $d > 3$ yielding a complete method in theory, but the restricted case of $d = 2, 3$ will be of the greatest practical interest.

4.2.3 Computation of the Hopf Condition in Convex Coordinates

The central task of this approach is to formulate a condition for the existence of Hopf bifurcations for each computed subsystem using convex coordinates and based on the condition for the existence of Hopf bifurcations with empty unstable manifolds. We first compute the Jacobian in reaction space using convex parameters, if the Jacobian is singular, we reduce the subsystem to the invariant manifold, we compute then a semi-algebraic formula expressing the condition for the occurrence of Hopf bifurcations, and finally, we generate the first-order existentially quantified formula.

4.2.3.1 Computation of the Jacobian in Reaction Space

As already demonstrated in Sect. 3.3, the transformation of the jacobian from concentration space into reaction space yields the following equation:

$$\text{Jac}(x) = \widehat{\text{Jac}}(j)\text{diag}(1/x_1, ..., 1/x_m), \qquad (4.10)$$

with $\widehat{\text{Jac}}(j) = \mathcal{S}\text{diag}(\sum_i j_i \mathcal{E}_i)\mathcal{K}^t$.

4.2.3.2 Jacobian on the Reduced Manifold

Conservation laws in a (bio)-chemical reaction system give rise to a singularity of the Jacobian of the entire polynomial system that presents the whole (bio)-chemical system and also of some Jacobian matrices of the computed subsystems. To compute the Hopf condition the Jacobian matrices should be transformed into nonsingular matrices. Therefore, we reduce them by computing the Jacobian Jac_i on the reduced manifolds using the method presented in sect. 2.1.2.

4.2.3.3 Semi-Algebraic Description of Hopf Bifurcations

We compute the Hopf condition based on the Hurwitz-Hopf criterion. Therefore, we compute the Hurwitz matrix and the Hurwitz determinants Δ_i. The Hopf condition of a subsystem can be expressed in reaction space using the semi-algebraic description shown in [23] by the following first-order formula:

$$\exists x(a_n > 0 \land \Delta_{n-1}(j,x) = 0 \land \Delta_{n-2}(j,x) > 0 \land \cdots \land \Delta_1(j,x) > 0) \qquad (4.11)$$

where n denotes the number of species in the reaction network.

Our method then involves the solution of these existentially quantified formulae, which can be computed using general packages for quantifier elimination on real closed fields yielding an answer of true or false, or packages to test for the satisfiability of the existentially quantified formulae yielding an answer of *satisfiable* (sat) or *unsatisfiable* (unsat).

4.2.4 Integration of Computational Logic Tools

We integrated into our computations all the computational logic tools listed in Sect. 2.2, which are all capable of solving formula (4.11). However, in this work, we present only results obtained with the freely available tools REDLOG and Z3, which provided the best computation time. REDLOG returns *true* and Z3 returns *sat* if the condition for the occurrence of a Hopf bifurcation is satisfied. If the condition is not satisfied, they return *false* and *unsat*, respectively. Because of the modular structure of our approach, we will be able to integrate other packages—either elements of commercial systems or novel developments—easily.

4.2.5 Pseudo-Code of the *HoCoQ* Algorithm

Alg. 1 summarizes the steps discussed above and outlines our method $HoCoQ$ in an algorithmic fashion.

Algorithm 1: $HoCoQ$ Method for Computing Hopf Bifurcations in Reaction Space.

Input: A chemical reaction network $\mathcal{N}$ with $\dim(\mathcal{N}) = n$.

Output: The algorithm returns a statement concerning the existence of a Hopf bifurcation

1 **begin**
2   R:= false;
3   generate the stoichiometric matrix $\mathcal{S}$ and kinetic matrix $\mathcal{K}$ from the reaction network
4   compute the minimal set $\mathcal{E}$ of the vectors generating the flux cone
5   **for** $d = 1 \ldots n$ **do**
6   compute all d-faces (subsystems) $\{\mathcal{N}_i\}_i$ of the flux cone
7   **for** *each subsystem* $\mathcal{N}_i$ **do**
8   compute from $\mathcal{K}$, $\mathcal{S}$ the transformed Jacobian Jac_i of $\mathcal{N}_i$ in terms of convex coordinates j_i;
9   **if** Jac_i *is singular* **then**
10   compute the reduced manifold of Jac_i calling the result also Jac_i
11   compute the characteristic polynomial χ_i of Jac_i;
12   compute the Hurwitz determinants of χ_i;
13   compute the Hopf existence condition for $\mathcal{N}_i$;
14   generate the first-order existentially quantified formula $\mathcal{F}_i$ expressing the Hopf existence condition, the constraints on the concentrations and the constraints on the cone coordinates;
15   reduce and simplify the generated formula $\mathcal{F}_i$
16   R:= $R \vee \mathcal{F}_i$
17   **return** R

4.2.6 Computation of Examples using $HoCoQ$ Method

We have applied our algorithm $HoCoQ$ on various chemical reaction networks that have been discussed in various monographs and for which the existing algorithms for the symbolic computations approach fails. We were able to detect the existence of Hopf bifurcations in some of them, which are listed below. We thereby demonstrate the results provided by REDLOG and Z3.

4.2.6.1 Phosphofructokinase Reaction

As a first example, we consider the main example used in the hand computation presented in [30]—the phosphofructokinase reaction. There are 3 chemical species and 7 reactions. S_1 denotes the product Fructose-1,6-biphosphate, S_2 denotes the reactant Fructose-6-phosphate, and the extension S_3 stands for another intermediate that is in equilibrium with Fructose-1,6-biphosphate. The network (4.12) represents the phosphofructokinase reaction.

$$2S_1 + S_2 \xrightarrow{k_1} 3S_1$$
$$S_2 \underset{k_4}{\overset{k_5}{\rightleftharpoons}} 0 \underset{k_3}{\overset{k_2}{\rightleftharpoons}} S_1 \underset{k_7}{\overset{k_6}{\rightleftharpoons}} S_3. \qquad (4.12)$$

This chemical reaction system yields the following stoichiometric matrix $\mathcal{S}_1$ and kinetic matrix $\mathcal{K}_1$:

$$\mathcal{S}_1 = \begin{pmatrix} 1 & 1 & -1 & 0 & 0 & -1 & 1 \\ -1 & 0 & 0 & 1 & -1 & 0 & 0 \\ 0 & 0 & 0 & 0 & 0 & 1 & -1 \end{pmatrix}$$

$$\mathcal{K}_1 = \begin{pmatrix} 2 & 0 & 1 & 0 & 0 & 1 & 0 \\ 1 & 0 & 0 & 0 & 1 & 0 & 0 \\ 0 & 0 & 0 & 0 & 0 & 0 & 1 \end{pmatrix}$$

The flux cone is spanned by the following four vectors (extreme currents):

$$\mathcal{E}_1 = \begin{pmatrix} 0 & 1 & 1 & 0 & 0 & 0 & 0 \end{pmatrix},$$
$$\mathcal{E}_2 = \begin{pmatrix} 0 & 0 & 0 & 1 & 1 & 0 & 0 \end{pmatrix},$$
$$\mathcal{E}_3 = \begin{pmatrix} 0 & 0 & 0 & 0 & 0 & 1 & 1 \end{pmatrix},$$
$$\mathcal{E}_4 = \begin{pmatrix} 1 & 0 & 1 & 1 & 0 & 0 & 0 \end{pmatrix}.$$

This problem has previously been investigated using its formulation in reaction coordinates in [80]. Using currently available quantifier elimination packages, the problem could not be solved in its parametric form. Only when using existential closure on the parameters could it be shown by successful quantifier eliminations performed in REDLOG that there exist positive parameters for which there exists a Hopf bifurcation fixed point in the positive orthant. When replicating the experiments we found that the situation described in [80] still applies.

The results on the subsystems involving 1-faces, 2-faces, 3-faces, and 4-faces are summarized in Table 4.1. A Hopf bifurcation can be found using the 1-face $\mathcal{E}_4$ and most of the subsystems extending it in less than one second. While Z3 provides no results for the 4-face $\mathcal{E}_1\mathcal{E}_2\mathcal{E}_3\mathcal{E}_4$ after 10000 seconds computation time, REDLOG requires only a few seconds of computation time to find a Hopf bifurcation fixed point.

TABLE 4.1: Computation of Hopf bifurcations in the phosphofructokinase reaction using *HoCoQ* algorithm

Subsystem	Redlog		Z3	
	Result	Time(s)	Result	Time(s)
$\mathcal{E}_1$	false	< 1	unsat	< 1
$\mathcal{E}_2$	false	< 1	unsat	< 1
$\mathcal{E}_3$	false	< 1	unsat	< 1
$\mathcal{E}_4$	true	< 1	sat	< 1
$\mathcal{E}_1\mathcal{E}_2$	false	< 1	unsat	< 1
$\mathcal{E}_1\mathcal{E}_3$	false	< 1	unsat	< 1
$\mathcal{E}_1\mathcal{E}_4$	true	< 1	sat	< 1
$\mathcal{E}_2\mathcal{E}_3$	false	< 1	unsat	< 1
$\mathcal{E}_2\mathcal{E}_4$	true	< 1	sat	< 1
$\mathcal{E}_3\mathcal{E}_4$	true	< 1	sat	< 1
$\mathcal{E}_1\mathcal{E}_2\mathcal{E}_3$	false	< 1	unsat	< 1
$\mathcal{E}_1\mathcal{E}_2\mathcal{E}_4$	true	< 1	sat	< 1
$\mathcal{E}_1\mathcal{E}_3\mathcal{E}_4$	true	1	sat	< 1
$\mathcal{E}_2\mathcal{E}_3\mathcal{E}_4$	true	2.5	sat	< 1
$\mathcal{E}_1\mathcal{E}_2\mathcal{E}_3\mathcal{E}_4$	true	6	no result	> 10000

4.2.6.2 Enzymatic Transfer of Calcium Ions

Our second example is a biochemical model that was investigated in [30]—the enzymatic transfer of calcium ions, Ca^{++}, across cellmembranes. It includes as shown in network (4.13) six reactions and four species, where S_1 stands for cytosolic Ca^{++}, S_2 stands for Ca^{++} in the endoplasmic reticulum, S_3 denotes the enzyme catalyzing the transport of Ca^{++} into the endoplasmic reticulum, and S_4 denotes the enzyme-substrate complex. This system is autocatalytic insofar as the concentration of cytosolic Ca^{++} stimulates the release of stored Ca^{++} from the endoplasmic reticulum [30].

$$0 \underset{k_{21}}{\overset{k_{12}}{\rightleftarrows}} S_1$$
$$S_1 + S_2 \xrightarrow{k_{43}} 2S_1$$
$$S_1 + S_3 \underset{k_{65}}{\overset{k_{56}}{\rightleftarrows}} S_4 \xrightarrow{k_{76}} S_2 + S_3 \tag{4.13}$$

The following stoichiometric matrix $\mathcal{S}_2$ and kinetic matrix $\mathcal{K}_2$ represent the kinetic description of the network (4.13).

$$\mathcal{S}_2 = \begin{pmatrix} -1 & 1 & 1 & 1 & -1 & 0 \\ 0 & 0 & -1 & 0 & 0 & 1 \\ 0 & 0 & 0 & 1 & -1 & 1 \\ 0 & 0 & 0 & -1 & 1 & -1 \end{pmatrix}$$

$$\mathcal{K}_2 = \begin{pmatrix} 1 & 0 & 1 & 0 & 1 & 0 \\ 0 & 0 & 1 & 0 & 0 & 0 \\ 0 & 0 & 0 & 0 & 1 & 0 \\ 0 & 0 & 0 & 1 & 0 & 1 \end{pmatrix}$$

$$\mathcal{E}_1 = \begin{pmatrix} 1 & 1 & 0 & 0 & 0 & 0 \end{pmatrix},$$
$$\mathcal{E}_2 = \begin{pmatrix} 0 & 0 & 1 & 0 & 1 & 1 \end{pmatrix},$$
$$\mathcal{E}_3 = \begin{pmatrix} 0 & 0 & 0 & 1 & 1 & 0 \end{pmatrix}.$$

For this system the Jacobian matrix is singular—therefore, in the classical sense there are no Hopf bifurcations. However, in the reduced system we find that there are Hopf bifurcations—and we can compute them in concentration space as well as using convex coordinates. The results and computation times are summarized in Table 4.2.

Chapter 4. *Detection of Hopf Bifurcation Using Convex Coordinates*   38

TABLE 4.2: Computation of Hopf bifurcations in the model "enzymatic transfer of calcium ions" using *HoCoQ* algorithm

Subsystem	Redlog		Z3	
	Result	Time(s)	Result	Time(s)
$\mathcal{E}_1$	false	< 1	unsat	< 1
$\mathcal{E}_2$	false	< 1	unsat	< 1
$\mathcal{E}_3$	false	< 1	unsat	< 1
$\mathcal{E}_1\mathcal{E}_2$	true	< 1	sat	< 1
$\mathcal{E}_1\mathcal{E}_3$	false	< 1	unsat	< 1
$\mathcal{E}_2\mathcal{E}_3$	false	< 1	unsat	< 1
$\mathcal{E}_1\mathcal{E}_2\mathcal{E}_3$	true	11	no result	> 10000

4.2.6.3 Model of Calcium Oscillations in the Cilia of Olfactory Sensory Neurons

As the next example, we consider the model for calcium oscillations in the cilia of olfactory sensory neurons discussed in [64]. The underlying mechanism of this model is based on direct negative regulation of cyclic nucleotide-gated channels by calcium/calmodulin and does not require any autocatalysis such as calcium-induced calcium release. Reidl et al. presented a mathematical model for this example in [64] and gave predictions for the parameter ranges in which oscillations should be observable. This model contains a fractional exponent ε, as shown in the following differential equations.

$$\frac{d}{dt}x = k_1 - k_5 x z$$
$$\frac{d}{dt}y = k_2 x - 4k_3 y^2 + 4k_4 z - k_6 y^\varepsilon$$
$$\frac{d}{dt}z = k_3 y^2 - k_4 z$$

The model yields the following stoichiometric matrix $\mathcal{S}_3$ and kinetic matrix $\mathcal{K}_3$:

$$\mathcal{S}_3 = \begin{pmatrix} 1 & 0 & 0 & 0 & -1 & 0 \\ 0 & 1 & -4 & 4 & 0 & -1 \\ 0 & 0 & 1 & -1 & 0 & 0 \end{pmatrix}$$

Chapter 4. Detection of Hopf Bifurcation Using Convex Coordinates

$$\mathcal{K}_3 = \begin{pmatrix} 0 & 1 & 0 & 0 & 1 & 0 \\ 0 & 0 & 2 & 0 & 0 & \varepsilon \\ 0 & 0 & 0 & 1 & 1 & 0 \end{pmatrix}$$

The representative vectors of the flux cone of this model are:

$$\begin{aligned} \mathcal{E}_1 &= \begin{pmatrix} 0 & 1 & 0 & 0 & 0 & 1 \end{pmatrix}, \\ \mathcal{E}_2 &= \begin{pmatrix} 0 & 0 & 1 & 1 & 0 & 0 \end{pmatrix}, \\ \mathcal{E}_3 &= \begin{pmatrix} 1 & 0 & 0 & 0 & 1 & 0 \end{pmatrix}. \end{aligned}$$

In concentration space the solution of a quantifier elimination problem is valid only for integer values of the parameter ε; this is because ε appears in the exponent, and the techniques of quantifier elimination over the ordered field of the reals is restricted to polynomials (or rational functions).

However, in the formulation in reaction coordinates the parameter ε appears as a variable with values in the real closed field used in the computations.

Therefore for a given subsystem we cannot ask only whether a Hopf bifurcation fixed point exists, but we can formulate the question with a free parameter ε.

The answer—a quantifier free formula involving ε—gives the condition for ε for which a Hopf bifurcation occurs for the subsystem. When investigating subsystems resulting from 2-faces we found no Hopf bifurcations, but for the parametric question on 3-faces we obtained the following answer in less than 10sec of computation time using a combination of REDLOG and QEPCAD:

$$\varepsilon + 2 > 0 \land 4\varepsilon - 1 < 0$$

Thus for $\varepsilon \in (-2, 0.25)$ we have shown that Hopf bifurcation fixed points exist (for suitable reaction constants). Using numerical simulations of this model Reidl et al. [64] could not find Hopf bifurcations for values of the parameter ε larger than approximately 0.05.

The Table 4.3 summarizes the results obtaining for 2- and 3- faces.

TABLE 4.3: Computation of Hopf bifurcations in model of calcium Oscillations using $HoCoQ$ algorithm

Sub-system	Redlog		Z3	
	Result	Time(s)	Result	Time(s)
$\mathcal{E}_1$	false	< 1	unsat	< 1
$\mathcal{E}_2$	false	< 1	unsat	< 1
$\mathcal{E}_3$	false	< 1	unsat	< 1
$\mathcal{E}_1\mathcal{E}_2$	false	< 1	unsat	< 1
$\mathcal{E}_1\mathcal{E}_3$	false	< 1	unsat	< 1
$\mathcal{E}_2\mathcal{E}_3$	false	< 1	unsat	< 1
$\mathcal{E}_1\mathcal{E}_2\mathcal{E}_3$	true	< 1	sat	< 1

4.3 $HoCaT$: Algorithm for Computing Hopf Bifurcations using Convex Coordinates and Tropical Geometry

The algorithmic method $HoCoQ$ discussed in Sect. 4.2 enabled us to determine the existence of Hopf bifurcations in various (bio)-chemical reaction networks even for those with conservation laws. For some chemical networks with complex dynamics, however, it remained difficult to process the final obtained quantified formulae with the currently available quantifier elimination packages.

In this section we present an efficient algorithmic approach, called $HoCaT$, which is sketched in Fig. 4.2. This algorithm uses the basic ideas of the previous algorithm $HoCoQ$, namely stoichiometric network analysis and manifold reduction method for systems with conservation laws. However, when the discussion provided in Sect. 4.1.1 for a criterion for the occurrence of Hopf bifurcations without requiring empty unstable manifolds is carried over to convex coordinates, the new condition for the existence of Hopf bifurcations is given by $\Delta_{n-1}(j,x) = 0$ only. Solving such single equations enables us to refrain from utilizing quantifier elimination techniques. Instead, the main algorithmic problem is to determine whether a single multivariate polynomial has a zero for positive coordinates.

For this purpose, in Sect. 4.3.1, we provide heuristics on the basis of the Newton polytope that ensure the existence of positive and negative values of the polynomial for positive coordinates, in Sect. 4.3.2, we present a summary of the $HoCaT$ Algorithm, and in Sect. 4.3.3 we apply our method to several (bio)-chemical reaction networks.

Chapter 4. Detection of Hopf Bifurcation Using Convex Coordinates

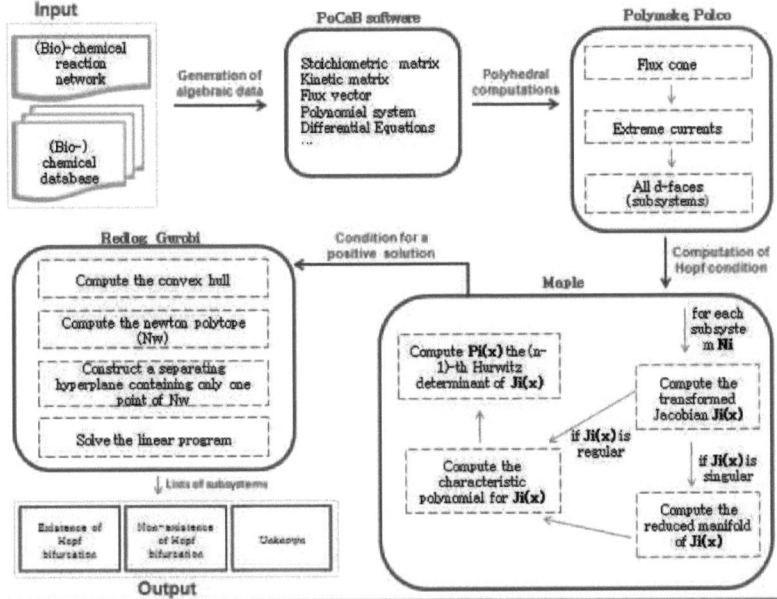

FIGURE 4.2: *HoCaT* method

4.3.1 Sufficient Conditions for a Positive Solution of a Single Multivariate Polynomial Equation

The method discussed in this section is summarized in an algorithmic way in Alg. 2, which uses Alg. 3 as a subalgorithm.

Given $f \in \mathbb{Z}[x_1, \ldots, x_m]$, our goal is to heuristically certify the existence of at least one zero $(z_1, \ldots, z_m) \in \,]0, \infty[^m$ for which all coordinates are strictly positive. To start with, we evaluate $f(1, \ldots, 1) = f_1 \in \mathbb{R}$. If $f_1 = 0$, then we are done. If $f_1 < 0$, then by the intermediate value theorem, it is sufficient to find $p \in \,]0, \infty[^m$ such that $f(p) > 0$. Similarly, if $f_1 > 0$ it is sufficient to find $p \in \,]0, \infty[^m$ such that $(-f)(p) > 0$. This algorithmically reduces our original problem to finding, for given $g \in \mathbb{Z}[x_1, \ldots, x_m]$, at least one $p \in \,]0, \infty[^m$ such that $g(p) = f_2 > 0$.

Algorithm 2: pzerop

Input: $f \in \mathbb{Z}[x_1, \ldots, x_m]$

Output: One of the following:

(A) 1, which means that $f(1, \ldots, 1) = 0$.

(B) (π, ν), where $\nu = (p, f(p))$ and $\pi = (q, f(q))$ for $p, q \in]0, \infty[^m$, which means that $f(p) < 0 < f(q)$. Then there is a zero on $]0, \infty[^m$ by the intermediate value theorem.

(C) +, which means that f has been identified as positive definite on $]0, \infty[^m$. Then there is no zero on $]0, \infty[^m$.

(D) −, which means that f has been identified as negative definite on $]0, \infty[^m$. Then there is no zero on $]0, \infty[^m$.

(E) $\bot$, which means that this incomplete procedure failed.

```
 1  begin
 2      f₁ := f(1, ..., 1)
 3      if f₁ = 0 then
 4          return 1
 5      else if f₁ < 0 then
 6          π := pzerop₁(f)
 7          ν := ((1, ..., 1), f₁)
 8          if π ∈ {⊥, −} then
 9              return π
10          else
11              return (ν, π)
12      else
13          π := ((1, ..., 1), f₁)
14          ν' := pzerop₁(−f)
15          if ν' = ⊥ then
16              return ⊥
17          else if ν' = − then
18              return +
19          else
20              (p, f(p)) := ν'
21              ν := (p, −f(p))
22              return (ν, π)
```

Chapter 4. Detection of Hopf Bifurcation Using Convex Coordinates

Algorithm 3: pzerop$_1$

Input: $f \in \mathbb{Z}[x_1, \ldots, x_m]$

Output: One of the following:

(A) $\pi = (q, f(q))$, where $q \in \;]0, \infty[^m$ with $0 < f(q)$.

(B) $-$, which means that f has been identified as negative definite on $]0, \infty[^m$. Then there is no zero on $]0, \infty[^m$.

(C) $\bot$, which means that this incomplete procedure failed.

```
 1  begin
 2      F⁺ := { d ∈ frame(f) | sgn(d) = 1 }
 3      if F⁺ = ∅ then
 4          return −
 5      foreach (d₁, ..., dₘ) ∈ F⁺ do
 6          L := {d₁n₁ + ··· + dₘnₘ − c = 0}
 7          foreach (e₁, ..., eₘ) ∈ frame(f) \ F⁺ do
 8              L := L ∪ {e₁n₁ + ··· + eₘnₘ − c ≤ −1}
 9          if L is feasible with solution (n₁, ..., nₘ, c) ∈ ℚ^{m+1} then
10              g := the principal denominator of n₁, ..., nₘ
11              (N₁, ..., Nₘ) := (gn₁, ..., gnₘ) ∈ ℤ^m
12              f̄ := f[x₁ ← ω^{N₁}, ..., xₘ ← ω^{Nₘ}] ∈ ℤ(ω)
13              assert lc(f̄) > 0 when using non-exact arithmetic in the LP solver
14              k := min{ k ∈ ℕ | f̄(2^k) > 0 }
15              return ((2^{kN₁}, ..., 2^{kNₘ}), f̄(2^k))
16      return ⊥
```

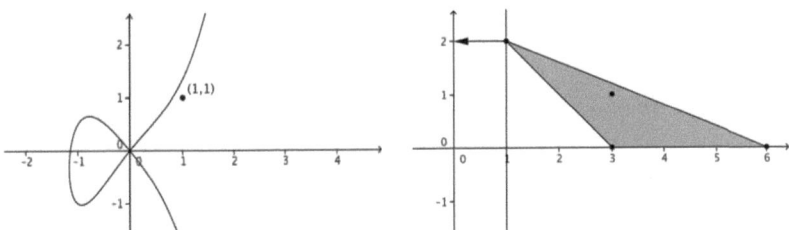

FIGURE 4.3: We consider $g_0 = -2x_1^6 + x_1^3 x_2 - 3x_1^3 + 2x_1 x_2^2$. The left hand shows the variety $g_0 = 0$. The right hand side shows the frame, the Newton polytope, and a separating hyperplane for the positive monomial $2x_1 x_2^2$ with its normal vector.

We will accompany the description of our method with the example $g_0 = -2x_1^6 + x_1^3 x_2 - 3x_1^3 + 2x_1 x_2^2 \in \mathbb{Z}[x_1, x_2]$. Fig. 4.3 shows an implicit plot of this polynomial. In addition to its variety, g_0 has three sign invariant regions, one bounded one and two unbounded ones. One of the unbounded regions contains our initial test point $(1, 1)$, for which we find that $g_0(1,1) = -2 < 0$. Therefore our goal is to find one point $p \in \,]0,\infty[^2$ such that $g_0(p) > 0$.

In the spirit of tropical geometry—and we refer to [81] as a standard reference with respect to its application for polynomial system solving—we take an abstract view of

$$g = \sum_{d \in D} a_d x^d := \sum_{(d_1,\ldots,d_m) \in D} a_{d_1,\ldots,d_m} x_1^{d_1} \cdots x_m^{d_m}$$

as the set $\mathrm{frame}(g) = D \subseteq \mathbb{N}^m$ of all exponent vectors of the contained monomials. For each $d \in \mathrm{frame}(g)$, we are able to determine $\mathrm{sgn}(d) := \mathrm{sgn}(a_d) \in \{-1, 1\}$. The set of vertices of the convex hull of the frame is called the *Newton polytope* $\mathrm{newton}(g) \subseteq \mathrm{frame}(g)$. In fact, the existence of at least one point $d^* \in \mathrm{newton}(g)$ with $\mathrm{sgn}(d^*) = 1$ is sufficient for the existence of $p \in \,]0,\infty[^m$ with $g(p) > 0$.

In our example, we have $\mathrm{frame}(g_0) = \{(6,0), (3,1), (3,0), (1,2)\}$ and $\mathrm{newton}(g_0) = \{(6,0), (3,0), (1,2)\} \subseteq \mathrm{frame}(g_0)$. We are particularly interested in $d^* = (d_1^*, d_2^*) = (1,2)$, which is the only point that has a positive sign as it corresponds to the monomial $2x_1 x_2^2$.

To understand this sufficient condition, we are now going to compute from d^* and g a suitable point p. We construct a hyperplane $H : n^T x = c$ containing d^* such that all other points of $\mathrm{newton}(g)$ are not contained in H and lie on the same side of H. We choose the normal vector $n \in \mathbb{R}^m$ such that it points into the halfspace that does not contain the Newton polytope. The vector $c \in \mathbb{R}^m$ is such that $\frac{c}{|n|}$ is the offset of H from the origin in the direction of n.

In our example H is the line $x = 1$ given by $n = (-1, 0)$ and $c = -1$. Fig. 4.3 depicts the situation.

Considering the standard scalar product $\langle \cdot | \cdot \rangle$, it turns out that generally $\langle n | d^* \rangle = \max\{ \langle n | d \rangle \mid d \in \mathrm{newton}(g) \}$, and that this maximum is strict. For the monomials of the original polynomial $g = \sum_{d \in D} a_d x^d$ and a new variable ω this observation translates via the following identity:

$$\bar{g} = g[x \leftarrow \omega^n] = \sum_{d \in D} a_d \omega^{\langle n | d \rangle} \in \mathbb{Z}(\omega).$$

Therefore, plugging a number $\beta \in \mathbb{R}$ into $\bar{g}$ corresponds to plugging the point $\beta^n \in \mathbb{R}^m$ into g and from our identity, we see that in $\bar{g}$ the exponent $\langle n | d^* \rangle$ corresponding to our chosen point $d^* \in \mathrm{newton}(g)$ dominates all other exponents, so for large β, the sign of $\bar{g}(\beta) = g(\beta^n)$ equals the positive sign of the coefficient a_{d^*} of the corresponding monomial. To find a suitable β, we successively compute $\bar{g}(2^k)$ for increasing $k \in \mathbb{N}$.

Chapter 4. Detection of Hopf Bifurcation Using Convex Coordinates 45

In our example we obtain $\bar{g} = 2\omega^{-1} - 2\omega^{-3} - 2\omega^{-6}$, and we obtain $\bar{g}(1) = -2$, but already $\bar{g}(2) = \frac{23}{32} > 0$. In terms of the original g this corresponds to plugging in the point $p = 2^{(-1,0)} = \left(\frac{1}{2}, 1\right) \in {]0,\infty[}^2$.

It remains to be clarified how to construct the hyperplane H. Consider frame$(g) = \{(d_{i1}, \ldots, d_{im}) \in \mathbb{N}^m \mid i \in \{1, \ldots, k\}\}$. If sgn$(d) = -1$ for all $d \in$ frame(g), then we know that g is negative definite on ${]0,\infty[}^m$. Otherwise, assume, without loss of generality, that sgn$(d_{11}, \ldots, d_{1m}) = 1$. We write down the following linear program:

$$\begin{pmatrix} d_{11} & \cdots & d_{1m} & -1 \end{pmatrix} \cdot \begin{pmatrix} n_1 \\ \vdots \\ n_m \\ c \end{pmatrix} = 0, \quad \begin{pmatrix} d_{21} & \cdots & d_{2m} & -1 \\ \vdots & \ddots & \vdots & \vdots \\ d_{k1} & \cdots & d_{km} & -1 \end{pmatrix} \cdot \begin{pmatrix} n_1 \\ \vdots \\ n_m \\ c \end{pmatrix} \leq -1.$$

This is feasible if and only if $(d_{11}, \ldots, d_{1m}) \in$ newton(g). In the negative case, we know that $(d_{11}, \ldots, d_{1m}) \in$ frame$(g) \setminus$ newton(g), and we iterate with another $d \in$ frame(g) with sgn$(d) = 1$. If we finally fail on all such d, then our incomplete algorithm has failed. In the positive case, the solution provides a normal vector $n = (n_1, \ldots, n_m)$ and the offset c for a suitable hyperplane H. Our linear program can be solved with any standard LP solver. For our purposes here, we have used Gurobi; the dual simplex of GLPSOL[3] also performs quite similarly on the input considered here.

For our example $g_0 = -2x_1^6 + x_1^3 x_2 - 3x_1^3 + 2x_1 x_2^2$, we generate the linear program

$$\begin{aligned} n_1 + 2n_2 - c &= 0 \\ 6n_1 - c &\leq -1 \\ 3n_1 + n_2 - c &\leq -1 \\ 3n_1 - c &\leq -1, \end{aligned}$$

for which Gurobi computes the solution $n = (n_1, n_2) = (-0.5, 0)$, $c = -0.5$. Notice that the solutions obtained from the LP solvers are typically floats, which we lift to integer vectors by suitable rounding and GCD computations.

Note that we do not explicitly construct the convex hull newton(g) of the frame(g) although there are advanced algorithms and implementations like QuickHull[4] available for this purpose. Instead we favour a linear programming approach for several reasons. Firstly, we do not require that comprehensive information, instead, it is sufficient to find one vertex of the covex hull that has a positive sign. Secondly, for the application dicussed here, it turns out that there typically exist only a few (approximately 10%) such

---
[3] www.gnu.org/software/glpk
[4] www.qhull.org

candidate points. Finally, it is known that for high dimensions, the subset of frame(g) establishing vertices of the convex hull gets comparatively large. Practical experiments using QuickHull on our data support these theoretical considerations.

4.3.2 Summarizing the *HoCaT* Algorithm

The steps involving the pre-precessing procedure, polyhedral computation, and computation of the reduced Jacobian that we previously used for the *HoCoQ* method and discussed in Sect. 4.2 remain the same. After computing the characteristic polynomial of the Jacobian matrix of each subsystem, we compute the $(n-1)^{\text{th}}$ Hurwitz determinant of the characteristic polynomial, and we apply Alg. 2 to check for positive solutions of the respective polynomial equations $\Delta_{n-1}(j,x) = 0$. Alg. 4 outlines our efficient approach in an algorithmic fashion.

4.3.3 Computation of Examples Using the *HoCaT* Method

In this section, we will demonstrate the efficiency of our novel approach *HoCaT* by analyzing several chemical networks with different dimensions. We will first compute Hopf bifurcations in the reaction networks already discussed in Sect. 4.2.6 using the *HoCaT* method. We will also wish to discuss and detect the occurrence of Hopf bifurcations in higher dimensional networks. We will therefore apply our new method to the 5-dimensional system of electro-oxidation of methanol presented in [66], to the well-known 9-dimensional example *MAPK* discussed in [13] and in other papers and to the 22-dimensional network modeling the control of DNA replication in fission yeast [53]. We will also compute Hopf bifurcations in the family of original models that describe a gene regulated by a polymer of its own protein, which are well-studied using the quasi-steady state approximation method in [4].

4.3.3.1 Phosphofructokinase Reaction

As the first example we consider the phosphofructokinase reaction discussed in 4.2.6.1. As shown in Table 4.4, using the *HoCaT* algorithm, we were able to detect the occurrence of Hopf bifurcations in less than 1 second for all computed faces. For comparison, in the case of 4-faces the *HoCoQ* method requires 6 seconds.

Chapter 4. Detection of Hopf Bifurcation Using Convex Coordinates   47

Algorithm 4: $HoCaT$ Method for Computing Hopf Bifurcations in Reaction Space.

Input: A chemical reaction network $\mathcal{N}$ with $\dim(\mathcal{N}) = n$.

Output: (L_t, L_f, L_u), which are defined as follows: L_t is a list of subsystems containing a Hopf bifurcation, L_f is a list of subsystems in which the occurrence of Hopf bifurcations is excluded, and L_u is a list of subsystems for which the incomplete sub-procedure pzerop fails.

1 **begin**
2     $L_t = \emptyset$
3     $L_f = \emptyset$
4     $L_u = \emptyset$
5   generate the stoichiometric matrix $\mathcal{S}$ and the kinetic matrix $\mathcal{K}$ of $\mathcal{N}$
6   compute the minimal set $\mathcal{E}$ of the vectors generating the flux cone
7   **for** $d = 1 \ldots n$ **do**
8   compute all d-faces (subsystems) $\{\mathcal{N}_i\}_i$ of the flux cone
9   **for** *each subsystem* $\mathcal{N}_i$ **do**
10   compute from $\mathcal{K}, \mathcal{S}$ the transformed Jacobian Jac_i of $\mathcal{N}_i$ in terms of convex coordinates j_i
11   **if** Jac_i *is singular* **then**
12   compute the reduced manifold of Jac_i calling the result also Jac_i
13   compute the characteristic polynomial χ_i of Jac_i
14   compute the $(n-1)^{\text{th}}$ Hurwitz determinant Δ_{n-1} of χ_i
15   compute $\mathcal{F}_i := \text{pzerop}(\Delta_{n-1}(j,x))$ using Algorithm 2
16   **if** $\mathcal{F}_i = 1$ or $\mathcal{F}_i$ *is of the form* (π, ν) **then**
17           $L_t := L_t \cup \{\mathcal{N}_i\}$
18   **else if** $\mathcal{F}_i = +$ or $\mathcal{F}_i = -$ **then**
19           $L_f := L_f \cup \{\mathcal{N}_i\}$
20   **else if** $\mathcal{F}_i = \bot$ **then**
21           $L_u := L_u \cup \{\mathcal{N}_i\}$
22   **return** (L_t, L_f, L_u)

4.3.3.2 Enzymatic Transfer of Calcium Ions

The computation of Hopf bifurcations in the model of the enzymatic transfer of calcium ions discussed in Sect. 4.2.6.2 using the $HoCaT$ method yields the results presented in Table 4.5.

While the $HoCoQ$ method requires 11 seconds of computation time for the 3-faces, the $HoCaT$ method needs less than 1 second.

Chapter 4. *Detection of Hopf Bifurcation Using Convex Coordinates*   48

TABLE 4.4: Computation of Hopf bifurcations in the phosphofructokinase reaction using *HoCaT* algorithm

Subsystem	Result	Time
$\mathcal{E}_1$	unsat	<1
$\mathcal{E}_2$	unsat	<1
$\mathcal{E}_3$	unsat	<1
$\mathcal{E}_4$	sat	<1
$\mathcal{E}_1\mathcal{E}_2$	unsat	<1
$\mathcal{E}_1\mathcal{E}_3$	unsat	<1
$\mathcal{E}_1\mathcal{E}_4$	sat	<1
$\mathcal{E}_2\mathcal{E}_3$	unsat	<1
$\mathcal{E}_2\mathcal{E}_4$	sat	<1
$\mathcal{E}_3\mathcal{E}_4$	sat	<1
$\mathcal{E}_1\mathcal{E}_2\mathcal{E}_3$	unsat	<1
$\mathcal{E}_1\mathcal{E}_2\mathcal{E}_4$	sat	<1
$\mathcal{E}_1\mathcal{E}_3\mathcal{E}_4$	sat	<1
$\mathcal{E}_2\mathcal{E}_3\mathcal{E}_4$	sat	<1
$\mathcal{E}_1\mathcal{E}_2\mathcal{E}_3\mathcal{E}_4$	sat	<1

TABLE 4.5: Computation of Hopf bifurcations in the model "enzymatic transfer of calcium ions" using *HoCaT* algorithm

Subsystem	Result	Time(s)
$\mathcal{E}_1$	unsat	<1
$\mathcal{E}_2$	unsat	<1
$\mathcal{E}_3$	unsat	<1
$\mathcal{E}_1\mathcal{E}_2$	sat	<1
$\mathcal{E}_1\mathcal{E}_3$	unsat	<1
$\mathcal{E}_2\mathcal{E}_3$	unsat	<1
$\mathcal{E}_1\mathcal{E}_2\mathcal{E}_3$	sat	<1

4.3.3.3   Model of Calcium Oscillations in the Cilia of Olfactory Sensory Neurons

Table 4.6 shows the results of computing Hopf bifurcations in the model calcium oscillations in the cilia of olfactory sensory neurons discussed in Sect. 4.2.6.3.

TABLE 4.6: Computation of Hopf bifurcations in the model "calcium oscillations in the cilia of olfactory sensory neurons" using *HoCaT* algorithm

Subsystem	Result	Time
$\mathcal{E}_1$	unsat	<1
$\mathcal{E}_2$	unsat	<1
$\mathcal{E}_3$	unsat	<1
$\mathcal{E}_1\mathcal{E}_2$	unsat	<1
$\mathcal{E}_1\mathcal{E}_3$	unsat	<1
$\mathcal{E}_2\mathcal{E}_3$	unsat	<1
$\mathcal{E}_1\mathcal{E}_2\mathcal{E}_3$	sat	<1

Chapter 4. Detection of Hopf Bifurcation Using Convex Coordinates   49

4.3.3.4 Electro-Oxidation of Methanol

Sauerbrei et al. [66] developed a model for a mechanism for the kinetic instabilities observed in the galvanostatic electro-oxidation of methanol. To keep the model simple, they neglected the side reactions and assumed that the whole process runs through HCO and CO. They then proposed the reaction network (4.14), which involves five essential species (nonessential species are enclosed in square brackets).

$$[\text{MeOH}_b] + 3* \xrightarrow{k_1, \Phi} \text{HCO} + [3\text{H}^+] + 3\text{e}^-$$
$$\text{HCO} \xrightarrow{k_2} \text{CO} + 2* + [\text{H}^+] + (\text{e}^-)$$
$$[\text{H}_2\text{O}] + * \xrightarrow{k_3, \Phi} \text{O} + [2\text{H}^+] + (2\text{e}^-)$$
$$\text{CO} + \text{O} \xrightarrow{k_4} 2* + [\text{CO}_2]$$
$$[2\text{H}^+] + (2\text{e}^-) + \text{O} \xrightarrow{k_5, -\Phi} * + [\text{H}_2\text{O}]. \tag{4.14}$$

Electrochemical reactions depend exponentially on the double layer potential Φ, so there is no power law kinetics initially. The system can, however, be transformed into power laws forms by using $x_3 = e^{k_6 \Phi}$ as a variable. By performing certain substitutions as shown in [66] the model yields the following differential equations and matrices. Note that this model has a negative exponent.

$$\begin{aligned}
\dot{x}_1 &= -3k_1 x_1^2 x_3 + 2k_2 x_4 - k_3 x_1 x_3 + 2k_4 x_2 x_5 + k_5 x_2 x_3^{-1} \\
\dot{x}_2 &= k_3 x_1 x_3 - k_4 x_2 x_5 - k_5 x_2 x_3^{-1} \\
\dot{x}_3 &= k_6 k_7 x_3 - k_1 k_6 x_1^2 x_3^2 \\
\dot{x}_4 &= k_1 x_1^2 x_3 - k_2 x_4 \\
\dot{x}_5 &= k_2 x_4 - k_4 k_2 x_5
\end{aligned} \tag{4.15}$$

$$S_4 = \begin{pmatrix} -3 & 2 & -1 & 2 & 1 & 0 & 0 \\ 0 & 0 & 1 & -1 & -1 & 0 & 0 \\ 0 & 0 & 0 & 0 & 0 & -1 & 1 \\ 1 & -1 & 0 & 0 & 0 & 0 & 0 \\ 0 & 1 & 0 & -1 & 0 & 0 & 0 \end{pmatrix}$$

$$\mathcal{K}_4 = \begin{pmatrix} 2 & 0 & 1 & 0 & 0 & 2 & 0 \\ 0 & 0 & 0 & 1 & 1 & 0 & 0 \\ 1 & 0 & 1 & 0 & -1 & 2 & 1 \\ 0 & 1 & 0 & 0 & 0 & 0 & 0 \\ 0 & 0 & 0 & 1 & 0 & 0 & 0 \end{pmatrix}$$

The stoichiometric matrix $\mathcal{S}_4$ yields the following extreme currents:

$$\mathcal{E}_1 = \begin{pmatrix} 0 & 0 & 1 & 0 & 1 & 0 & 0 \end{pmatrix},$$
$$\mathcal{E}_2 = \begin{pmatrix} 1 & 1 & 1 & 1 & 0 & 0 & 0 \end{pmatrix},$$
$$\mathcal{E}_3 = \begin{pmatrix} 0 & 0 & 0 & 0 & 0 & 1 & 1 \end{pmatrix}.$$

We applied the *HoCaT* algorithm to all possible faces and we were able to find the occurrence of Hopf bifurcations in the 2-faces $\mathcal{E}_2\mathcal{E}_3$ and the 3-faces $\mathcal{E}_1\mathcal{E}_2\mathcal{E}_3$ as shown in Table 4.8.

TABLE 4.7: Computation of Hopf bifurcations in the model "electro-oxidation of methanol" using *HoCaT* algorithm

Subsystem	Result	Time(s)
$\mathcal{E}_1$	unsat	< 1
$\mathcal{E}_2$	unsat	< 1
$\mathcal{E}_3$	unsat	< 1
$\mathcal{E}_1\mathcal{E}_2$	unsat	< 1
$\mathcal{E}_1\mathcal{E}_3$	unsat	< 1
$\mathcal{E}_2\mathcal{E}_3$	sat	< 1
$\mathcal{E}_1\mathcal{E}_2\mathcal{E}_3$	sat	< 1

4.3.3.5 Methylene Blue Oscillator System

As the next example we apply the *HoCaT* method on the well-known complex autocatalytic *methylen blue oscillator (MBO)* system. We attempted to compute Hopf bifurcations in all subsystems of this model that involve 2-faces and 3-faces using our original *HoCoQ* approach, but the generated quantified formulae could not be solved by quantifier elimination, even with main memory of up to 500 GB and computation times of up to one week. The *MBO* model is described by the reaction network (4.16):

Chapter 4. Detection of Hopf Bifurcation Using Convex Coordinates

$$\begin{aligned}
MB^+ + HS^- &\longrightarrow MB + HS \\
H_2O + MB + HS^- &\longrightarrow MBH + HS + OH^- \\
HS + OH^- + MB^+ &\longrightarrow MB + S + H_2O \\
H_2O + 2MB &\longrightarrow MB^+ + MBH + OH^- \\
HS^- + O_2 &\longrightarrow HS + O_2^- \\
HS + O_2 + OH^- &\longrightarrow O_2^- + S + H_2O \\
2H_2O + HS^- + O_2^- &\longrightarrow H_2O_2 + HS + 2OH^- \\
O_2^- + HS + H_2O &\longrightarrow H_2O_2 + S + H_2O \\
H_2O_2 + 2HS^- &\longrightarrow 2HS + 2OH^- \\
MB + O_2 &\longrightarrow MB^+ + O_2^- \\
HS^- + MB + H_2O_2 &\longrightarrow MB^+ + HS + 2OH^- \\
OH^- + 2HS &\longrightarrow HS^- + S + H_2O \\
MB + HS &\longrightarrow MBH + S \\
H_2O + MBH + O_2^- &\longrightarrow MB + H_2O_2 + OH^- \\
&\longrightarrow O_2 \quad\quad\quad\quad (4.16)
\end{aligned}$$

The MBO reaction system contains 15 reactions and 11 species O_2, O_2^-, HS, MB^+, MB, MBH, HS^-, OH^-, S, and H_2O_2. It may be reduced to a six dimensional system by considering only the essential species MB, MB^+, HS, MBH, O_2, and O_2^-.

The pre-processing step of our algorithm yields the following two matrices describing the reaction laws: stoichiometric matrix $\mathcal{S}$ and kinetic matrix $\mathcal{K}$.

$$\mathcal{S}_5 = \begin{pmatrix}
1 & -1 & 1 & -2 & 0 & 0 & 0 & 0 & 0 & -1 & -1 & 0 & -1 & 1 & 0 \\
-1 & 0 & -1 & 1 & 0 & 0 & 0 & 0 & 0 & 1 & 1 & 0 & 0 & 0 & 0 \\
1 & 1 & -1 & 0 & 1 & -1 & 1 & -1 & 2 & 0 & 1 & -2 & -1 & 0 & 0 \\
0 & 1 & 0 & 1 & 0 & 0 & 0 & 0 & 0 & 0 & 0 & 0 & 1 & -1 & 0 \\
0 & 0 & 0 & 0 & -1 & -1 & 0 & 0 & 0 & -1 & 0 & 0 & 0 & 0 & 1 \\
0 & 0 & 0 & 0 & 1 & 1 & -1 & -1 & 0 & 1 & 0 & 0 & 0 & -1 & 0
\end{pmatrix}$$

$$\mathcal{K}_5 = \begin{pmatrix}
0 & 1 & 0 & 2 & 0 & 0 & 0 & 0 & 0 & 1 & 1 & 0 & 1 & 0 & 0 \\
1 & 0 & 1 & 0 & 0 & 0 & 0 & 0 & 0 & 0 & 0 & 0 & 0 & 0 & 0 \\
0 & 0 & 1 & 0 & 0 & 1 & 0 & 1 & 0 & 0 & 0 & 2 & 1 & 0 & 0 \\
0 & 0 & 0 & 0 & 0 & 0 & 0 & 0 & 0 & 0 & 0 & 0 & 0 & 1 & 0 \\
0 & 0 & 0 & 0 & 1 & 1 & 0 & 0 & 0 & 1 & 0 & 0 & 0 & 0 & 0 \\
0 & 0 & 0 & 0 & 0 & 0 & 1 & 1 & 0 & 0 & 0 & 0 & 0 & 1 & 0
\end{pmatrix}.$$

The flux cone of this model is spanned by 28 extreme currents. There are 187 subsystems of 2-faces and 549 subsystems of 3-faces. Using our new approach $HoCaT$ we were able to detect Hopf bifurcations in the 1-face subsystem generated by the extreme current

$$\mathcal{E} = (0\ \ 0\ \ 1\ \ 0\ \ 0\ \ 0\ \ 0\ \ 0\ \ 1\ \ 1\ \ 0\ \ 0\ \ 1\ \ 1\ \ 1)$$

and in 105 cases of 2-faces. The following table summarize the results.

TABLE 4.8: Results of the computation of Hopf bifurcations in 1-face and 2- faces using $HoCaT$

Subsystems	Number of cases	Satisfied	Unsatisfied	Unknown
1-face	28	1	27	0
2-faces	187	105	66	15

All computations on a single instance required at most 350 milliseconds of CPU time.

Recall that a positive answer for at least one of the cases guarantees the existence of a Hopf bifurcation for the original system in spite of the fact that there are cases without a definite answer.

4.3.3.6 Mitogen-Activated Protein Kinase ($MAPK$)

We next consider a well-studied model in cell biology that describes the activity of mitogen-activated protein kinase ($MAPK$). This model is known to exhibit bistability, namely it has up to two stable equilibria, if the parameter vector is located in an appropriate region of parameter space [19, 50]. Conradi et al. also studied this model in [13] and mentioned that finding these regions, for example by using numerical tools like bifurcation analysis, is a non-trivial task as it amounts to searching the entire parameter space. They show that for a model of a single layer of a $MAPK$ cascade it is possible to derive analytical descriptions of these regions through the use of mass action kinetics. As an example, we compute Hopf bifurcations in the extensively studied 9-dimensional network (4.17) that belongs to a family of network structures that has been postulated as a model for a single layer of a $MAPK$ cascade. We use here the same notations as in [13]. We use A as a placeholder for either $MAPKK$ or a $MAPK$, E_1 for mono-phosphorylated $MAPKKK$ or double-phosphorylated $MAPKK$, and E_2 for $MAPKK$'ase or $MAPK$'ase.

Chapter 4. *Detection of Hopf Bifurcation Using Convex Coordinates*

$$A + E_1 \underset{k_2}{\overset{k_1}{\rightleftharpoons}} AE_1 \overset{k_3}{\rightarrow} A_p + E_1 \underset{k_5}{\overset{k_4}{\rightleftharpoons}} A_pE_1 \overset{k_6}{\rightarrow} A_{pp} + E_1,$$

$$A_{pp} + E2 \underset{k_8}{\overset{k_7}{\rightleftharpoons}} A_{pp}E_2 \overset{k_9}{\rightarrow} A_p + E_2 \underset{k_{11}}{\overset{k_{10}}{\rightleftharpoons}} A_pE_2 \overset{k_{12}}{\rightarrow} A + E_2. \qquad (4.17)$$

The *MAPK* network (4.17) involves twelve reactions and nine species, A, E_1, AE_1, A_p, A_pE_1, A_{pp}, $E2$, $A_{pp}E_2$, and A_pE_2. The appropriate stoichiometric matrix $\mathcal{S}_6$ and kinetic matrix $\mathcal{K}_6$ are as follows:

$$\mathcal{S}_6 = \begin{pmatrix} -1 & 1 & 0 & 0 & 0 & 0 & 0 & 0 & 0 & 0 & 0 & 1 \\ -1 & 1 & 1 & -1 & 1 & 1 & 0 & 0 & 0 & 0 & 0 & 0 \\ 1 & -1 & -1 & 0 & 0 & 0 & 0 & 0 & 0 & 0 & 0 & 0 \\ 0 & 0 & 1 & -1 & 1 & 0 & 0 & 0 & 1 & -1 & 1 & 0 \\ 0 & 0 & 0 & 1 & -1 & -1 & 0 & 0 & 0 & 0 & 0 & 0 \\ 0 & 0 & 0 & 0 & 0 & 1 & -1 & 1 & 0 & 0 & 0 & 0 \\ 0 & 0 & 0 & 0 & 0 & 0 & -1 & 1 & 1 & -1 & 1 & 1 \\ 0 & 0 & 0 & 0 & 0 & 0 & 1 & -1 & -1 & 0 & 0 & 0 \\ 0 & 0 & 0 & 0 & 0 & 0 & 0 & 0 & 0 & 1 & -1 & -1 \end{pmatrix}$$

$$\mathcal{K}_6 = \begin{pmatrix} 1 & 0 & 0 & 0 & 0 & 0 & 0 & 0 & 0 & 0 & 0 & 0 \\ 1 & 0 & 0 & 1 & 0 & 0 & 0 & 0 & 0 & 0 & 0 & 0 \\ 0 & 1 & 1 & 0 & 0 & 0 & 0 & 0 & 0 & 0 & 0 & 0 \\ 0 & 0 & 0 & 1 & 0 & 0 & 0 & 0 & 1 & 0 & 0 & 0 \\ 0 & 0 & 0 & 0 & 1 & 1 & 0 & 0 & 0 & 0 & 0 & 0 \\ 0 & 0 & 0 & 0 & 0 & 0 & 1 & 0 & 0 & 0 & 0 & 0 \\ 0 & 0 & 0 & 0 & 0 & 0 & 1 & 0 & 0 & 1 & 0 & 0 \\ 0 & 0 & 0 & 0 & 0 & 0 & 0 & 1 & 1 & 0 & 0 & 0 \\ 0 & 0 & 0 & 0 & 0 & 0 & 0 & 0 & 0 & 0 & 1 & 1 \end{pmatrix}.$$

The flux cone of the *MAPK* network is spanned by the following six vectors of extreme currents:

$$\mathcal{E}_1 = \begin{pmatrix} 1 & 1 & 0 & 0 & 0 & 0 & 0 & 0 & 0 & 0 & 0 & 0 \end{pmatrix},$$
$$\mathcal{E}_2 = \begin{pmatrix} 0 & 0 & 0 & 1 & 1 & 0 & 0 & 0 & 0 & 0 & 0 & 0 \end{pmatrix},$$
$$\mathcal{E}_3 = \begin{pmatrix} 0 & 0 & 0 & 0 & 0 & 0 & 1 & 1 & 0 & 0 & 0 & 0 \end{pmatrix},$$
$$\mathcal{E}_4 = \begin{pmatrix} 0 & 0 & 0 & 0 & 0 & 0 & 0 & 0 & 1 & 1 & 0 \end{pmatrix},$$
$$\mathcal{E}_5 = \begin{pmatrix} 1 & 0 & 1 & 0 & 0 & 0 & 0 & 0 & 1 & 0 & 1 \end{pmatrix},$$
$$\mathcal{E}_6 = \begin{pmatrix} 0 & 0 & 0 & 1 & 0 & 1 & 1 & 0 & 1 & 0 & 0 & 0 \end{pmatrix}.$$

Although it is difficult to compute Hopf bifurcations in the $MAPK$ networks, we were able to detect the occurrence of a Hopf bifurcation using our algorithm in the subsystem generated by the 2-face of $\mathcal{E}_5$ and $\mathcal{E}_6$ in 25 seconds of computation time. For all the subsystems generated by 1-faces or by other 2-faces we could exclude them.

4.3.3.7 Models of Genetic Circuits

Boulier et al. [4] studied the use of a rigorous quasi-steady state approximation method to determine the existence of Hopf bifurcations in a family of models describing a gene regulated by a polymer of its own protein. This family of models is dependent on an integer parameter n that expresses the number of polymerizations and on featuring a negative feedback loop. The model sketched in Fig. 4.4 [5] describes a single gene regulated by a polymer that is obtained by combining a protein n times. The variables G and H represent the state of the gene. The mRNA concentration and the concentration of the protein translated from the mRNA are represented by M and P, respectively. The n types of polymers of P are denoted by $G = P_1, P_2, \ldots, P_n$. Greek letters represent parameters [4].

The models of genetic ciruits yields the following reaction laws.

$$G + P_n \underset{\theta}{\overset{\alpha}{\rightleftharpoons}} H, \quad G \xrightarrow{\rho_f} G + M, \quad H \xrightarrow{\rho_b} H + M,$$

$$M \xrightarrow{\beta} M + P, \quad M \xrightarrow{\delta_M} \emptyset, \quad P \xrightarrow{\delta_P} \emptyset, \quad P_i + P \underset{k_i^-}{\overset{k_i^+}{\rightleftharpoons}} P_{i+1} \quad (1 \leq i \leq n-1). \quad (4.18)$$

Applying a rigorous quasi-steady state approximation and several rescalings of the variables and parameters yields the following family of ordinary differential equations [4]:

---
[5]Figure from [4]

Chapter 4. Detection of Hopf Bifurcation Using Convex Coordinates

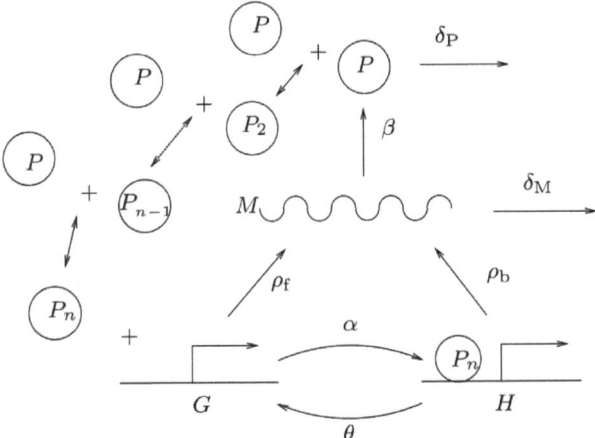

FIGURE 4.4: A gene regulated by a polymer of its protein

$$\frac{\mathrm{d}}{\mathrm{d}t}G(t) = \theta(\gamma_0 - G(t) - G(t)P(t)^n),$$
$$\frac{\mathrm{d}}{\mathrm{d}t}P(t) = n\alpha(\gamma_0 - G(t) - G(t)P(t)^n) + \delta(M(t) - P(t)),$$
$$\frac{\mathrm{d}}{\mathrm{d}t}M(t) = \lambda_1 G(t) + \gamma_0 \mu - M(t), \tag{4.19}$$

where n is a natural number.

Sturm et al. [79, 80] also analyzed the existence of Hopf bifurcations in the 3-dimensional steady-state approximation of the models shown in (4.19). They computed its occurrence in concentration space up to $n = 10$ and they found the absence of Hopf bifurcations in the family of models for $n \leq 8$ and its existence for $n \geq 9$.

We investigated the existence of Hopf bifurcations in the original family of models for $n = 2, \ldots, 10$, wherein we also considered the fast reactions. Each model thus involved then $3 + n$ species and yields corresponding to the stoichiometric matrix and kinetic matrix. The number of the vectors that span the flux cone is dependent on the parameter n, which expresses the number of polymerizations and effect that for increasing n. We applied our *HoCaT* method to all 9 models and in contrast to the results of the quasi-steady state method, we were able to detect the existence of Hopf bifurcations for $n \geq 3$ and its absence for $n = 2$.

To elucidate the cause of the occurrence of Hopf bifurcations for $n \geq 3$ in the original state of the systems, we carefully analyzed the results of the system with $n = 3$

polymerizations. The system yields the following stoichiometric and kinetic matrices:

$$\mathcal{S}_7 = \begin{pmatrix} -1 & 1 & 0 & 0 & 0 & 0 & 0 & 0 & 0 & 0 & 0 \\ 1 & -1 & 0 & 0 & 0 & 0 & 0 & 0 & 0 & 0 & 0 \\ 0 & 0 & 1 & 1 & 0 & -1 & 0 & 0 & 0 & 0 & 0 \\ 0 & 0 & 0 & 0 & 1 & 0 & -1 & -2 & 2 & -1 & 1 \\ 0 & 0 & 0 & 0 & 0 & 0 & 0 & 1 & -1 & -1 & 1 \\ -1 & 1 & 0 & 0 & 0 & 0 & 0 & 0 & 0 & 1 & -1 \end{pmatrix}$$

$$\mathcal{K}_7 = \begin{pmatrix} 1 & 0 & 1 & 0 & 0 & 0 & 0 & 0 & 0 & 0 & 0 \\ 0 & 1 & 0 & 1 & 0 & 0 & 0 & 0 & 0 & 0 & 0 \\ 0 & 0 & 0 & 0 & 1 & 1 & 0 & 0 & 0 & 0 & 0 \\ 0 & 0 & 0 & 0 & 0 & 0 & 1 & 2 & 0 & 1 & 0 \\ 0 & 0 & 0 & 1 & 0 & 0 & 0 & 0 & 1 & 0 & 0 \\ 1 & 0 & 0 & 0 & 0 & 0 & 0 & 0 & 0 & 0 & 1 \end{pmatrix}.$$

The following six extreme currents represent the flux cone:

$$\mathcal{E}_1 = \begin{pmatrix} 0 & 0 & 0 & 0 & 1 & 0 & 1 & 0 & 0 & 0 & 0 \end{pmatrix},$$
$$\mathcal{E}_2 = \begin{pmatrix} 0 & 0 & 0 & 0 & 0 & 0 & 0 & 0 & 0 & 1 & 1 \end{pmatrix},$$
$$\mathcal{E}_3 = \begin{pmatrix} 1 & 1 & 0 & 0 & 0 & 0 & 0 & 0 & 0 & 0 & 0 \end{pmatrix},$$
$$\mathcal{E}_4 = \begin{pmatrix} 0 & 0 & 1 & 0 & 0 & 1 & 0 & 0 & 0 & 0 & 0 \end{pmatrix},$$
$$\mathcal{E}_5 = \begin{pmatrix} 0 & 0 & 0 & 1 & 0 & 1 & 0 & 0 & 0 & 0 & 0 \end{pmatrix},$$
$$\mathcal{E}_6 = \begin{pmatrix} 0 & 0 & 0 & 0 & 0 & 0 & 0 & 1 & 1 & 0 & 0 \end{pmatrix}.$$

We observed the absence of Hopf bifurcations in the 1-faces and 2-faces and its presence in one 3-face $\mathcal{E}_1\mathcal{E}_2\mathcal{E}_5$ generated by the vectors $\mathcal{E}_1,\mathcal{E}_2$, and $\mathcal{E}_5$, where $\mathcal{E}_5$ represents a reversible fast reaction. We also detected its existence in the trivial cases of 4-faces that contain the subsystem $\mathcal{E}_1\mathcal{E}_2\mathcal{E}_5$ and the subsystem $\mathcal{E}_1\mathcal{E}_3\mathcal{E}_5\mathcal{E}_6$, where $\mathcal{E}_6$ also represent a reversible fast reaction. We conclude that eliminating fast reactions in the system for quasi-steady state approximation causes the disappearance of Hopf bifurcations for $n \geq 3$.

4.3.3.8 Control of DNA Replication in Fission Yeast

As another high-dimensional example, we consider the 22 dimensional model that describes the control of DNA replication in fission yeast. It is described in [53] and stored as a curated model in the BioModels database (see Sect. 3.6.2.1) with the ID BIOMOD0007. The stoichiometric matrix, the kinetic matrix, the set of extreme currents, and other algebraic data for this example can be obtained from the database provided by our platform *PoCaB*. The flux cone of this model is spanned by 22 extreme currents and yields 231 2-faces and 1540 3-faces.

Using the *HoCaT* method we were able to detect the existence of Hopf bifurcations in 69 cases of the 3-faces and its absence in the 2-faces. The computation of this example also demonstrates the efficiency of our method, as it enables even the analysis of a 22-dimensional system.

Chapter 5

On Muldowney's Criteria for Polynomial Vector Fields with Constraints

In this chapter, we investigate the absence of global oscillations in polynomial systems by studying Muldowney's extension of the classical Bendixson-Dulac criterion for excluding periodic orbits to higher dimensions for polynomial vector fields. Using the formulation of Muldowney's sufficient criteria for excluding periodic orbits of the parameterized vector field on a convex set as a quantifier elimination problem over the ordered field of the reals we provide case studies of some systems arising in the life sciences. We discuss the use of simple conservation constraints and the use of parametric constraints for describing simple convex polytopes on which periodic orbits can be excluded by Muldowney's criteria.

5.1 Introduction and Preliminaries

In the study of ordinary differential equations the analysis of periodic trajectories is seen as an important goal in addition to describing the dynamics around fixed points. However, already for two-dimensional polynomial systems this question is related to Hilbert's 16th problem, which is still unsolved [42].

For the two-dimensional case the Bendixson-Dulac criterion gives a sufficient condition for the non-existence of periodic orbits. This criterion is parameterized by a Dulac function, and various techniques have been proposed to construct Dulac functions, which

Chapter 5. *On Muldowney's Criteria for Polynomial Vector Fields*   59

range form algebraic constructions for special systems to techniques involving the solution of certain partial differential equations [8–11, 55].

For the higher-dimensional case there are extensions of the criterion of Bendixson-Dulac that also allow the use of Dulac functions [51]. However, little work seems to have been done to construct Dulac functions in the higher dimensional cases—except for addressing it as a problem [92, 93].

Moreover, the common case of algebraic constraints in the simple form of conservation constraints have been used in ad hoc form by many authors—mainly to reduce 3D systems to 2D systems to be able to use the classical Bendixson-Dulac criterion—but have not been discussed in a more general setting.

In case studies of some systems arising in the life sciences we discuss the use of simple conservation constraints in a first line of investigation.

On the example of classical SIRS epidemiological model we show that even in this rather simple case different algorithmic strategies to use conservation constraints might lead to non-conclusive results for some, whereas others lead to conclusive results. Thus the fact that Muldowney's criteria are not coordinate independent pose an algorithmic problem.

We also discuss the use of parametric constraints for describing simple convex polytopes on which periodic orbits can be excluded by Muldowney's criteria. We will show that for a 3-dimensional model of viral dynamics [86], for which Muldwowney's criteria cannot exclude the existence of periodic orbits on the entire positive real octant, there is a cuboid on which periodic orbits can be excluded.

5.1.1 The Bendixson-Dulac Criterion for 2-Dimensional Vector Fields

Consider an autonomous planar vector field

$$\frac{dx}{dt} = F(x,y), \qquad \frac{dy}{dt} = G(x,y), \qquad (x,y) \in \mathbb{R}^2.$$

Bendixson in 1901 [2] was the first to give a criterion yielding sufficient conditions for excluding oscillations. Dulac in 1937 [22] was able to generalize the result of Bendixson as follows:

Theorem 5.1 (Bendixson-Dulac criterion)**.** *Let $B(x,y)$ be a scalar continuously differentiable function defined on a simply connected region $D \subset \mathbb{R}^2$ with no holes in it. If $\frac{\partial (BF)}{\partial x} + \frac{\partial (BG)}{\partial y}$ is not identically zero and does not change sign in D, then there are no periodic orbits lying entirely in D.*

For a modern proof we refer to [36, Theorem 1.8.2].

A common class of Dulac functions uses $B(x,y) = e^{(U(x,y))}$, see e.g. [11]. By the chain rule the exponential function can be factored out yielding $e^U \left(\frac{\partial U}{\partial x} F + \frac{\partial U}{\partial y} G + \frac{\partial F}{\partial x} + \frac{\partial G}{\partial y} \right)$. Hence, if $F, G, \frac{\partial U}{\partial x}$, and $\frac{\partial U}{\partial y}$ are rational functions, the Bendixson-Dulac criterion remains in the realm of the ordered field of the reals.

5.1.2 Muldowney's Extensions of the Bendixson-Dulac Criterion to Higher Dimensions

The algorithmic criteria discussed in the following can be seen as generalizations of the Bendixson-Dulac criterion for 2-dimensional vector fields to arbitrary dimensions.

The following theorem was proved by Muldowney [51, Theorem 4.1]: Suppose that one of the inequalities

$$\mu\left(\frac{\partial f^{[2]}}{\partial x}\right) < 0, \qquad \mu\left(-\frac{\partial f^{[2]}}{\partial x}\right) < 0 \qquad (5.1)$$

holds for all $x \in \mathbb{R}^n$. Then the autonomous system with vector field $f : \mathbb{R}^n \longrightarrow \mathbb{R}^n$ has no nonconstant periodic solutions. Here μ is some Lozinskiĭ norm and $f^{[2]}$ is one of the "compound matrices" of the Jacobian of the vector field f defined in [51]. As is also shown in [51] the criterion given in [51, Theorem 4.1] also holds when $x \in C$, where $C \subseteq \mathbb{R}^n$ is open and convex.

Remark. When $n = 2, \partial f^{[2]}/\partial x = \text{Trace} \, \partial f / \partial x = \text{div} f$, so that [51, Theorem 4.1] basically yield the results of Bendixson, i.e. the criterion of Muldowney can be seen as a generalization of the criterion of Bendixson from the planar case to arbitrary dimensions.

According to [51, (2.2)], the following expressions may be used as $\mu \left(\partial f^{[2]}/\partial x \right)$ in [51, Theorem 4.1], if the underlying norms for μ are the 1-norm, ∞-norm, and 2-norm respectively:

$$\max\left\{ \frac{\partial f_r}{\partial x_r} + \frac{\partial f_s}{\partial x_s} + \sum_{q \neq r,s} \left|\frac{\partial f_q}{\partial x_r}\right| + \left|\frac{\partial f_q}{\partial x_s}\right| : r,s = 1,\ldots,n, \, r \neq s \right\}, \qquad (5.2)$$

$$\max\left\{ \frac{\partial f_r}{\partial x_r} + \frac{\partial f_s}{\partial x_s} + \sum_{q \neq r,s} \left|\frac{\partial f_r}{\partial x_q}\right| + \left|\frac{\partial f_s}{\partial x_q}\right| : r,s = 1,\ldots,n, \, r \neq s \right\}, \qquad (5.3)$$

$$\lambda_1 + \lambda_2, \qquad (5.4)$$

where λ_1, λ_2 are the two largest eigenvalues of $\left(\partial f^*/\partial x + \partial f/\partial x\right)/2$.

Thus for a formula Γ over the reals defining an open convex subset C of $\mathbb{R}^n$ and an autonomous polynomial vector field $f : \mathbb{R}^n \to \mathbb{R}^n$ a first-order formula φ over the ordered

Chapter 5. On Muldowney's Criteria for Polynomial Vector Fields

field of the reals defines a sufficient condition such that the vector field defined by f has no non-constant periodic solution on C. As usual with real quantifier elimination we use the language of ordered rings. In addition, we admit function symbols for the maximum and for the absolute values, which are both definable.

Specifically, for the criterion involving the 1-norm we obtain

$$\varphi_1 \equiv \forall x_1 \forall x_2 \cdots \forall x_n \bigg(\Gamma \implies \qquad (5.5)$$
$$\max\bigg\{ \frac{\partial f_r}{\partial x_r} + \frac{\partial f_s}{\partial x_s} + \sum_{q \neq r,s} \bigg|\frac{\partial f_q}{\partial x_r}\bigg| + \bigg|\frac{\partial f_q}{\partial x_s}\bigg| : r,s = 1, \ldots, n,\ r \neq s \bigg\} < 0 \bigg),$$

and for the criterion involving the ∞-norm we obtain

$$\varphi_\infty \equiv \forall x_1 \forall x_2 \cdots \forall x_n \bigg(\Gamma \implies \qquad (5.6)$$
$$\max\bigg\{ \frac{\partial f_r}{\partial x_r} + \frac{\partial f_s}{\partial x_s} + \sum_{q \neq r,s} \bigg|\frac{\partial f_r}{\partial x_q}\bigg| + \bigg|\frac{\partial f_s}{\partial x_q}\bigg| : r,s = 1, \ldots, n,\ r \neq s \bigg\} < 0 \bigg).$$

In [92] the problem of efficient automatic resolution of maxima and absolute values is addressed and computation examples are given. If all variables and parameters are known to be positive, the technique of *positive quantifier elimination* can be used.

5.1.2.1 Extending Muldowney's Criteria with Dulac Functions.

Although a simple generalization of the Dulac criterion to higher dimensions does not seem to hold in the general setting [51], for *positive* functions $0 < r \in C^1(\mathbb{R}^n \longrightarrow \mathbb{R})$ one can replace f by rf in [51, Theorem 4.1], cf. (5.1). The rather simple proof is given in [51, Remark (d)].

If $B = e^U$ is used as a Dulac test function then by the chain rule the exponential function can be factored out also for Muldowney's criteria and the criterion remains in the realm of the ordered field of the reals, if all partial derivatives of U are rational functions.

5.1.2.2 Using Conservation Constraints.

Any algebraic constraints on the vector field can be transferred into the first-order formula over the ordered field of the reals expressing Muldowney's criteria. Simple conservation constraints stating that the sum of certain state variables is constant—conditions that are commonly found in chemical reaction systems or in epidemiological models—will not induce a failure of the degree limited virtual substitution methods [96] for quantifier elimination, if these were successful on the unconstrained system.

Nevertheless, an elimination of a variable by the others in a conservation constrained will reduce the dimension of the system and thus change Muldowney's criteria instead of adding another equality to Muldowney's criteria on the original system. We will report on the results of some systematic tests on the simple SIRS system in Sect. 5.2.1.

5.1.2.3 Parametric Specification of a Convex Subset.

The first-order formula γ specifying the convex subset on which a proof for the non-existence of periodic orbits is sought by Muldowney's criteria can very well contain parameters, too. The quantifier elimination procedure automatically yields conditions on the parameters that are exact with respect to Muldowney's criteria—potentially not mentioning input parameters if no constraint on any of them is necessary.

In Sect. 5.2.2 we will use this technique using simple parametric cuboids in a case, for which the Muldowney criteria do not give a conclusive answer on the entire positive real octant, but the specification of a 3-dimensional parametric cuboid shows that only a parametric restriction on one variable is necessary.

5.2 Case Studies

5.2.1 The SIRS Epidemiological Model

We consider the widely used SIRS epidemiological model, a parameterized formally 3-dimensional system of ordinary differential equations, cf. (5.7–5.9). The systems is widely used and well studied [7, 38, 46, 58, 94]. So we will not provide new insights into the structure of the system, but it is well suited as a test object for our algorithmic methods.

To account for the lost of immunity, the classical susceptible (S), infected (I) and recovered (R) model is adjusted by allowing a fraction of the recovered individuals R to move back into the susceptible pool S at a rate γ. This susceptible, infected, recovered and susceptible (SIRS) model is expressed as

$$\frac{d}{dt}S(t) = \mu\left(S(t) + I(t) + R(t)\right) - \mu S(t) - \beta S(t) I(t) + \gamma R(t) \tag{5.7}$$

$$\frac{d}{dt}I(t) = \beta S(t) I(t) - (\mu + \nu) I(t) \tag{5.8}$$

$$\frac{d}{dt}R(t) = \nu I(t) - (\mu + \gamma) R(t) \tag{5.9}$$

where ν is the rate of loss of infectiousness and the total population size N remains constant (i.e. $S + I + R = N$ is constant). The parameter μ represents both, the birth and mortality rates. Assuming that birth and mortality rates are equal is justified on the grounds that the annual infection rate is considerably higher than the population growth. The parameter β is the transmission rate of the infection.

5.2.1.1 Using Ad-hoc Reductions to 2D-Models

In the literature, reductions to 2D models using $S + I + R = N$ and replacing a suitable variable are commonly used. However, the question, which variable to choose is never addressed. In the following we give results for all possibilities showing that even for this simple example the results strongly differ. In all cases we use the scaling $N = 1$.

Eliminating R by $R = 1 - (I + S)$. In this case the criterion using the Dulac test function 1 returned the non-conclusive *true* as answer for $\neg\varphi$. However, using the Dulac function $\frac{1}{I(t)}$ the conclusive *false* as answer for $\neg\varphi$ was found within some milliseconds of computation time by REDLOG.

Eliminating I by $I = 1 - (S + R)$. Also in this case the criterion using the Dulac test function 1 returned the non-conclusive *true* as answer for $\neg\varphi$. We also obtained the the non-conclusive *true* as answer for $\neg\varphi$ when using the following Dulac test functions:

$$\frac{1}{R(t)S(t)}$$
$$\frac{1}{S(t)}$$
$$\frac{1}{R(t)}$$
$$R(t)$$
$$S(t)$$

Moreover, the computations using REDLOG did not come up with answers within 60 sec of computation time for several other Dulac test functions.

So using this elimination we did not come up with a conclusive answer by the Muldowney criteria.

Eliminating S by $S = 1-(I+R)$. In this case the criterion using the Dulac function 1 returned $\beta - \gamma - 2\mu - \nu > 0$ as answer for $\neg\varphi$. Using the Dulac function $\frac{1}{I(t)}$ returned the conclusive *false*, as was the case for the Dulac function $\frac{1}{I(t)R(t)}$; for the Dulac function $\frac{1}{R(t)}$ the criterion returned $\beta - \mu - \nu > 0$. As the conclusive *false* was found for some Dulac function, we thus have proved that the SIRS system does not have periodic orbits on the positive real octant.

5.2.1.2 Computations on the 3D-Model

Unconstrained model. For the 3D-SIRS model *not* using any conservation constraint the criterion using the Dulac test function 1 returned the non-conclusive *true* as answer for $\neg\varphi$. For all other Dulac tests functions we used we either obtained the non-conclusive *true* as answer for $\neg\varphi$, or REDLOG could not come up with a result within 60 sec of computation time.

5.2.2 A Model of Viral Dynamics

The following example is discussed in more depth in [92]. It consists of a simple mathematical model for the population dynamics of the human immunodefficiency type 1 virus (HIV-1) investigated in [86]. There a three-component model is described involving uninfected CD4 + T-cells, infected such cells and free viruses, whose densities at time t are denoted by $x(t), y(t), v(t)$, respectively.

$$\frac{d}{dt}x(t) = s - \mu x(t) - kx(t)y(t)$$
$$\frac{d}{dt}y(t) = kx(t)y(t) - \alpha y(t)$$

$$\frac{d}{dt}x(t) = s - \mu x(t) - \beta x(t)v(t)$$
$$\frac{d}{dt}y(t) = \beta x(t)v(t) - \alpha y(t)$$
$$\frac{d}{dt}v(t) = cy(t) - \gamma v(t)$$

FIGURE 5.1: The 2D- and 3D-Tuckwell-Wan examples

In [86] a simplified two-component model employed by Bonhoeffer et al. [3] is investigated analytically. In [86] using the general Bendixson-Dulac criteria for 2D-vector fields with an ad hoc Dulac function $B(x, y) = 1/y$ it is shown that there are no periodic

Chapter 5. On Muldowney's Criteria for Polynomial Vector Fields

TABLE 5.1: Results for the 2D-Tuckwall-Wan example (cf. Fig. 5.1) on the full positive octant

The computation times are the ones for the positive quantifier elimination in REDLOG.

Tuckwell-Wan 2D model			Used Dulac test function					
	1	$\frac{1}{x}$	$\frac{1}{y}$	$\frac{1}{xy}$	$\frac{1}{x+y}$	x	y	xy
Comp. Time [sec]	0.07	0.07	0.02	0.02	0.07	0.09	0.07	0.07
Result ($\neg\varphi$)	pc	pc	**false**	**false**	pc	pc	pc	pc

Here pc is the positivity condition on the parameters.

solutions for the system for positive parameter values and positive values of the state variables, i.e. the biologically relevant ones.

Remark. By "ad hoc" Dulac function we mean that the authors provide this function only and show that it is a Dulac function, but no other functions. No explanations or hints are given to the reader how this function was obtained.

In Table 5.1 the results for various low-degree rational and polynomial Dulac test functions are summarized. Notice that computation times for generating the formulas are negligible for these examples. Note that for $\neg\varphi$ the answer **false** gives the conclusive proof on the non-existence of periodic orbits on the positive cone.

As can be seen from the computation times given in Table 5.1 the quantifier elimination problems are not too hard. When performing tests with QEPCAD we could also solve all of these quantifier elimination problems in less than one second of computation time.

For the 3D-Tuckwell-Wan Model we tried several Dulac test functions but could not exclude the existence of a periodic orbit on $\mathbb{R}^{+3}$ for any of them. When specifying the parametric cube $(0, u_x) \times (0, u_y) \times (0, u_v)$ by adding the conditions $x(t) < u_x$, $y(t) < u_y$, and $v(t) < u_v$ for new parameters $u_x > 0$, $u_y > 0$, and $u_v > 0$—cf. Sect. 5.1.2.3—and using the trivial Dulac function 1—we obtain the following first-order formula for $\neg\varphi$ using Muldowney's criterion for the 1-norm (displayed in slightly hand edited form for better readability):

$$\exists v_1 \exists v_2 \exists v_3 : 0 < v_1 \wedge 0 < v_2 \wedge 0 < v_3 \wedge 0 < u_v \wedge 0 < u_x \wedge 0 < u_y \wedge$$
$$0 < c \wedge 0 < \mu \wedge 0 < s \wedge 0 < \alpha \wedge 0 < \beta \wedge 0 < \gamma \wedge$$
$$v_1 < u_v \wedge v_2 < u_x \wedge v_3 < u_y \wedge$$
$$0 \leq \max(-\gamma - \alpha + |\beta v_2|, -\mu - \beta v_1 - \alpha + |c|, -\gamma - \mu - \beta v_1 + |\beta v_2| + |\beta v_1|)$$

This quantifier elimination problem can also be solved "by hand" rather easily, and accordingly in less than 0.1 sec of computation time we obtain by the positive quantifier

Chapter 5. *On Muldowney's Criteria for Polynomial Vector Fields* 66

elimination procedure of REDLOG the following quantifier-free equivalent for φ:

$$\min\left(\frac{\alpha+\gamma}{\beta}, \frac{\mu+\gamma}{\beta}\right) \geq u_x \wedge \alpha + \mu \geq c \tag{5.10}$$

For better readability we have provided in (5.10) a slightly hand-edited version of the result formula.

Notice that there is no dependency on u_y and u_v, i.e. we have given a proof that the parametric 3D-Tuckwall-Wan does not have periodic orbits on

$$(0, u_x) \times (0, \infty) \times (0, \infty)$$

provided u_x (and $\alpha, \mu, \gamma, \beta$) fulfills the condition given in (5.10).

Chapter 6

Computing Stability in Convex Coordinates

The stability analysis may be used with some confidence to determine which chemical models are capable of exotic dynamics and thus the possibility of the existence of oscillations [12]. In Chapter 4, we described the $HoCoQ$ algorithm which improves the computation of Hopf bifurcations using stoichiometric network analysis and quantifier elimination. These ideas will be reused in a new algorithmic approach to investigate the stability of (bio)-chemical reaction networks. Our new algorithm for computing stability is similar to $HoCoQ$ except the condition for the existence of Hopf bifurcations. This will be replaced by the criteria which determine the stability of reaction systems. Therefore, we integrated two conditions which prove if all roots of a real polynomial are in the left half-plane of the complex plane, namely Hurwitz condition [41] and the Gantmacher condition that is based on Stieltjes theorem [28], and also the condition that checks if an extreme subnetwork is mixing stable. In this chapter, we introduce the mentioned conditions and we present the results obtained from applying our approach on some reaction networks already discussed in this book.

6.1 Computing Stability Using Hurwitz Criterion

A parameterized autonomous ordinary differential equation of the form $\dot{x} = f(u, x)$ with a scalar parameter u is asymptotically stable at the point (u_0, x_0), if $f(u_0, x_0) = 0$ and all eigenvalues of the Jacobian $D_x f(u_0, x_0)$ lie in the open left half of the complex plane. A well-known mathematical test to check this condition was introduced by Hurwitz in 1869

Chapter 6. *Computing Stability in Convex Coordinates* 68

[41]. He proved that a necessary and sufficient condition for all roots of the polynomial

$$p(\lambda) = a_0\lambda^n + a_1\lambda^{n-1} + \cdots + a_n \qquad (6.1)$$

have negative real part is that the values of determinants $\Delta_1, \Delta_2, \Delta_3, \ldots, \Delta_n$ are positive, where

$$\Delta_i = \begin{vmatrix} a_1 & a_3 & a_5 & \cdots & a_{2i-1} \\ a_0 & a_2 & a_4 & \cdots & a_{2i-2} \\ 0 & a_1 & a_3 & \cdots & a_{2i-3} \\ \cdots & \cdots & \cdots & \cdots & \cdots \\ \cdots & \cdots & \cdots & \cdots & a_i \end{vmatrix} \qquad (6.2)$$

are the Hurwitz determinants associated to the characteristic polynomial of the matrix $D_x f(u_0, x_0)$. In [23], it is shown that for a parameterized nonlinear system $\dot{x} = f(u, x)$, a natural question is to ask for which values of the parameter u the system is asymptotically stable near all its fixed points. This can be symbolically expressed by the first-order formula

$$\forall x (f(u, x) = 0 \implies \Delta_1(u, x) > 0, \ldots, \Delta_n(u, x) > 0). \qquad (6.3)$$

As shown in Sect. 3.3 the Jacobian matrix of a subsystem formed by d-faces is given by the following equation, where $\mathcal{S}$, $\mathcal{K}$ and $\mathcal{E}$ denote the stoichiometric matrix, kinetic matrix, and the set of extreme currents, respectively.

$$\text{Jac}(x) = \mathcal{S}\text{diag}(\sum_i^d j_i \mathcal{E}_i)\mathcal{K}^t \text{diag}(1/x_1, \ldots, 1/x_n).$$

For checking for asymptotically stability of a (bio)-chemical system near all its fixed points using convex coordinates we have to decide the satisfiability of the following formula:

$$\forall j_1 \cdots j_d \forall x_1 \cdots x_n (j_1 \geq 0 \land \cdots \land j_d \geq 0 \land x_1 \geq 0 \land \cdots \land x_n \geq 0$$
$$\implies \Delta_1(j, x) > 0, \ldots, \Delta_n(j, x) > 0).$$

For some (bio)-chemical networks already discussed in Chapter 4, namely phosphofructokinase reaction, model of enzymatic transfer of calcium ions, model for calcium oscillations in the cilia of olfactory sensory neurons, and model of electro-oxidation of

Chapter 6. Computing Stability in Convex Coordinates

methanol, we computed the stability in convex coordinates using Hurwitz condition. The following tables summarize the results provided by REDLOG and Z3.

TABLE 6.1: Computation of stability in the phosphofructokinase reaction in convex coordinates using Hurwitz condition

Subsystem	Redlog		Z3	
	Result	Time(s)	Result	Time(s)
$\mathcal{E}_1$	true	< 1	sat	< 1
$\mathcal{E}_2$	true	< 1	sat	< 1
$\mathcal{E}_3$	true	< 1	sat	< 1
$\mathcal{E}_4$	false	< 1	unsat	< 1
$\mathcal{E}_1\mathcal{E}_2$	true	< 1	unknown	< 1
$\mathcal{E}_1\mathcal{E}_3$	true	< 1	unknown	1.5
$\mathcal{E}_1\mathcal{E}_4$	false	< 1	unknown	2
$\mathcal{E}_2\mathcal{E}_3$	true	< 1	sat	< 1
$\mathcal{E}_2\mathcal{E}_4$	false	< 1	unsat	< 1
$\mathcal{E}_3\mathcal{E}_4$	false	< 1	unknown	7.5
$\mathcal{E}_1\mathcal{E}_2\mathcal{E}_3$	true	< 1	unknown	11
$\mathcal{E}_1\mathcal{E}_2\mathcal{E}_4$	false	< 1	unknown	< 1
$\mathcal{E}_1\mathcal{E}_3\mathcal{E}_4$	false	< 1	no result	> 10000
$\mathcal{E}_2\mathcal{E}_3\mathcal{E}_4$	false	< 1	no result	> 10000
$\mathcal{E}_1\mathcal{E}_2\mathcal{E}_3\mathcal{E}_4$	false	< 1	no result	> 10000

By comparing the REDLOG results of Hurwitz stability in Table 6.1 and the results of computing Hopf bifurcations in Table 4.4, we conclude that each unstable subsystem of the phosphofructokinase reaction undergoes a Hopf bifurcation.

TABLE 6.2: Computation of stability in the model "Enzymatic transfer of calcium ions" in convex coordinates using Hurwitz condition

Subsystem	Redlog		Z3	
	Result	Time(s)	Result	Time(s)
$\mathcal{E}_1$	true	< 1	sat	< 1
$\mathcal{E}_2$	false	< 1	unknown	2
$\mathcal{E}_3$	true	< 1	sat	< 1
$\mathcal{E}_1\mathcal{E}_2$	false	< 1	no result	> 10000
$\mathcal{E}_1\mathcal{E}_3$	true	< 1	sat	< 1
$\mathcal{E}_2\mathcal{E}_3$	false	< 1	unknown	13
$\mathcal{E}_1\mathcal{E}_2\mathcal{E}_3$	false	1.5	no result	> 1000

Chapter 6. *Computing Stability in Convex Coordinates* 70

For the model 'enzymatic transfer of calcium ions' we conclude by considering the results in Table 6.2 and in Table 4.5 that the absence of Hopf bifurcations occurs in two instable subsystems $\mathcal{E}_2$ and $\mathcal{E}_2\mathcal{E}_3$.

TABLE 6.3: Computation of stability in the model "calcium oscillations in the cilia of olfactory sensory neurons" using convex coordinates and Hurwitz condition

Subsystem	Redlog		Z3	
	Result	Time(s)	Result	Time(s)
$\mathcal{E}_1$	true*	< 1	sat	< 1
$\mathcal{E}_2$	true	< 1	sat	< 1
$\mathcal{E}_3$	true	< 1	sat	< 1
$\mathcal{E}_1\mathcal{E}_2$	true*	< 1	unknown	< 1
$\mathcal{E}_1\mathcal{E}_3$	true*	< 1	unknown	< 1
$\mathcal{E}_2\mathcal{E}_3$	true	< 1	sat	< 1
$\mathcal{E}_1\mathcal{E}_2\mathcal{E}_3$	no result	> 10000	no result	> 10000

The REDLOG results demonstrate that the subsystems $\mathcal{E}_1$, $\mathcal{E}_1\mathcal{E}_2$, and $\mathcal{E}_1\mathcal{E}_3$ of the model 'calcium oscillations in the cilia of olfactory sensory neurons' are stable, only if the exponent ε is positive (marked by true*). Hence all 1-face and 2-faces are stable and affirm our results in Sect. 4.2.6.3 Table 4.6 that they do not undergo Hopf bifurcations. The instability of subsystem $\mathcal{E}_1\mathcal{E}_2\mathcal{E}_3$ could not be computed even after 10000 seconds computation time.

TABLE 6.4: Computation of stability in the model "electro-oxidation of methanol" using convex coordinates and Hurwitz condition

Subsystem	Redlog		Z3	
	Result	Time(s)	Result	Time(s)
$\mathcal{E}_1$	true	< 1	sat	< 1
$\mathcal{E}_2$	false	< 1	no result	> 10000
$\mathcal{E}_3$	true	< 1	sat	< 1
$\mathcal{E}_1\mathcal{E}_2$	false	< 1	no result	> 10000
$\mathcal{E}_1\mathcal{E}_3$	false	< 1	unsat	< 1
$\mathcal{E}_2\mathcal{E}_3$	no result	> 10000	no result	> 10000
$\mathcal{E}_1\mathcal{E}_2\mathcal{E}_3$	no result	> 10000	no result	> 10000

We make the conclusion that the subsystems $\mathcal{E}_2$, $\mathcal{E}_1\mathcal{E}_2$, and $\mathcal{E}_2\mathcal{E}_3$ of the model 'electro-oxidation of methanol' are unstable but they do not undergo Hopf bifurcations. Z3 and also REDLOG are not able to compute the instability of the subsystems $\mathcal{E}_2\mathcal{E}_3$ and $\mathcal{E}_1\mathcal{E}_2\mathcal{E}_3$. As shown in Table 4.8, we found Hopf bifurcations in these two subsystems.

We also conclude from the above results that the computation of stability in convex coordinates using quantifier elimination is more efficient than computing stability in concentration space, but it is still a hard task to compute stability in 3-faces and even in 2-faces of (bio)-chemical networks with complex dynamic. The results also demonstrate the efficient of REDLOG in comparison to Z3.

6.2 Computing Stability Using Gantmacher-Stieltjes Criterion

We represent the real polynomial $p(\lambda)$ (6.1) in the form

$$p(\lambda) = h(\lambda^2) + \lambda g(\lambda^2). \tag{6.4}$$

Gantmacher introduced in [28] a condition for stability that is based on Stieltjes theorem and represent the Hurwitz polynomials by continued fractions. He investigated what conditions have to be imposed on $h(u)$ and $g(u)$ in order that $p(\lambda)$ be a Hurwitz polynomial, which means that all roots of $p(\lambda)$ are in the left half-plane of the complex plane.

The Stieltjes theorem says:

Theorem 6.1. *If $h(u)$, $g(u)$ is a positive pair o polynomials and $h(u)$ is of degree m, then*

$$\frac{g(u)}{h(u)} = c_0 + \cfrac{1}{d_0 u + \cfrac{1}{c_1 + \cfrac{1}{d_1 u + \cfrac{1}{c_2 + \cdots + \cfrac{1}{d_{m-1}u + \cfrac{1}{c_m}}}}}} \tag{6.5}$$

From Stieltjes theorem and from the observation that a polynomial $p(\lambda) = h(\lambda^2) + \lambda g(\lambda^2)$ is a Hurwitz polynomial if and only if $h(u)$ and $g(u)$ form a positive pair, Gantmacher deduced a criterion for stability, which is formulated in the following theorem:

Theorem 6.2. *A real polynomial of degree n $p(\lambda) = h(\lambda^2) + \lambda g(\lambda^2)$ is a Hurwitz polynomial if and only if the formula 6.5 holds with nonnegative c_0, and positive $c1, \cdots, c_m$, $d_0, \cdots, d_{m-1}$. Here $c_0 > 0$ when n is odd and $c_0 = 0$ when n is even.*

We replaced the Hurwitz criterion with Gantmacher-Stieltjes criterion in our approach for computing stability in convex coordinates and we again computed the stability in the networks discussed in previous section. Comparing the results provided by REDLOG of both methods, we conclude that both Hurwitz criterion and Gantmacher-Stieltjes criterion yield the same results. However REDLOG takes in various subsystems more time for simplifying the quantified formula that expresses the stability using Gantmacher criterion. The tables below contain in detail all results of treated networks.

TABLE 6.5: Computation of stability in the phosphofructokinase reaction using convex coordinates and Gantmacher-Stieltjes condition

Subsystem	Result	Time
$\mathcal{E}_1$	true	< 1
$\mathcal{E}_2$	true	< 1
$\mathcal{E}_3$	true	< 1
$\mathcal{E}_4$	false	< 1
$\mathcal{E}_1\mathcal{E}_2$	true	< 1
$\mathcal{E}_1\mathcal{E}_3$	true	< 1
$\mathcal{E}_1\mathcal{E}_4$	false	< 1
$\mathcal{E}_2\mathcal{E}_3$	true	< 1
$\mathcal{E}_2\mathcal{E}_4$	false	< 1
$\mathcal{E}_3\mathcal{E}_4$	false	< 1
$\mathcal{E}_1\mathcal{E}_2\mathcal{E}_3$	true	< 1
$\mathcal{E}_1\mathcal{E}_2\mathcal{E}_4$	false	< 1
$\mathcal{E}_1\mathcal{E}_3\mathcal{E}_4$	false	1
$\mathcal{E}_2\mathcal{E}_3\mathcal{E}_4$	false	1
$\mathcal{E}_1\mathcal{E}_2\mathcal{E}_3\mathcal{E}_4$	false	4

TABLE 6.6: Computation of stability in the model "enzymatic transfer of calcium ions" using convex coordinates and Gantmacher-Stieltjes condition

Subsystem	Result	Time(s)
$\mathcal{E}_1$	true	< 1
$\mathcal{E}_2$	false	< 1
$\mathcal{E}_3$	true	< 1
$\mathcal{E}_1\mathcal{E}_2$	false	< 1
$\mathcal{E}_1\mathcal{E}_3$	true	< 1
$\mathcal{E}_2\mathcal{E}_3$	false	< 1
$\mathcal{E}_1\mathcal{E}_2\mathcal{E}_3$	false	10000

TABLE 6.7: Computation of stability in the model "calcium oscillations in the cilia of olfactory sensory neurons" using convex coordinates and Gantmacher-Stieltjes condition

Subsystem	Result	Time
$\mathcal{E}_1$	true*	< 1
$\mathcal{E}_2$	true	< 1
$\mathcal{E}_3$	true	< 1
$\mathcal{E}_1\mathcal{E}_2$	true*	< 1
$\mathcal{E}_1\mathcal{E}_3$	true*	< 1
$\mathcal{E}_2\mathcal{E}_3$	true	< 1
$\mathcal{E}_1\mathcal{E}_2\mathcal{E}_3$	no result	> 10000

TABLE 6.8: Computation of stability in the model "electro-oxidation of methanol" using convex coordinates and Gantmacher-Stieltjes condition

Subsystem	Result	Time(s)
$\mathcal{E}_1$	true	< 1
$\mathcal{E}_2$	false	< 1
$\mathcal{E}_3$	true	< 1
$\mathcal{E}_1\mathcal{E}_2$	false	5
$\mathcal{E}_1\mathcal{E}_3$	false	< 1
$\mathcal{E}_2\mathcal{E}_3$	no result	> 10000
$\mathcal{E}_1\mathcal{E}_2\mathcal{E}_3$	no result	> 10000

6.3 Computation of Mixing Stability

Clarke invented in 1980 [12] a new term for computing stability of extreme subnetworks in convex coordinates. It is based on a special Lyapunov function and is called *mixing stability*. A network is an extreme network if it contains only one extreme current.

We consider the Jacobian matrix of an extreme subnetwork $\mathcal{N}_l$ that contains the extreme current $\mathcal{E}_l$:

$$\text{Jac}_l(x) = \mathcal{S}\text{diag}j_i\mathcal{E}_l\mathcal{K}^t\text{diag}(1/x_1,...,1/x_n). \tag{6.6}$$

The extreme subnetwork $\mathcal{N}_l$ is mixing stable if the symmetric matrix

$$\text{Jac}_s = \frac{\text{Jac}_l(x) + \text{Jac}_l{}^t(x)}{2} \tag{6.7}$$

Chapter 6. *Computing Stability in Convex Coordinates* 74

is *positive definite*, where $\text{Jac}_l{}^t(x)$ is the transpose of $\text{Jac}_l(x)$. The stability of a chemical network can be proved if it can be decomposed into mixing stable extreme networks. Clarke proved that if all extreme subnetworks are positive definite then all steady states for all parameter values are stable. Thereby, the mixing stability allows a network $\mathcal{N}_l$ to combine with any other mixing stable network and retain its stability. [12]

An extreme subnetwork may be stable and mixing stable, or stable and non-mixing stable, or unstable. An unstable network must contain at least one non-mixing stable extreme subnetwork. An extreme subnetwork which is stable and mixing stable, i.e which is stable and does not induce instability if it is linearly combined with other stable extreme subnetworks is called *positive circuit*. An unstable or non-mixing stable extreme current is called *stoichiometric generator* [12, 71]. For the (bio)-chemical networks discussed in the previous sections, we computed the mixing stability of all extreme subnetworks in convex coordinates using quantifier elimination. The tables below summarize the results provided by REDLOG.

For the phosphofructokinase reaction and the model 'Enzymatic transfer of calcium ions' (cf. Tables 6.9 and 6.10) we obtained the same results for the stability and mixing stability of extreme subnetworks. The stable subnetworks are also mixing stable. In the phosphofructokinase reaction there is only one non-mixing stable subnetwork $\mathcal{E}_4$ that also undergoes a Hopf-bifurcation (cf. Table 4.4). In the model 'Enzymatic transfer of calcium ions' the combination of the mixing stable subnetwork $\mathcal{E}_1$ and non-mixing stable subnetwork $\mathcal{E}_2$ gives rise to the occurrence of Hopf bifurcations in the subsystem $\mathcal{E}_1\mathcal{E}_2$ (cf. Table 4.5).

TABLE 6.9: Computation of mixing stability in the phosphofructokinase reaction

Subsystem	Result	Time
$\mathcal{E}_1$	true	< 1
$\mathcal{E}_2$	true	< 1
$\mathcal{E}_3$	true	< 1
$\mathcal{E}_4$	false	< 1

TABLE 6.10: Computation of mixing stability in the model "enzymatic transfer of calcium ions"

Subsystem	Result	Time(s)
$\mathcal{E}_1$	true	< 1
$\mathcal{E}_2$	false	< 1
$\mathcal{E}_3$	true	< 1

Chapter 6. *Computing Stability in Convex Coordinates* 75

As examples for models containing extreme subnetworks that are both stable and non-mixing stable, we consider two models 'Calcium Oscillations in the cilia of olfactory sensory neurons' and 'electro-oxidation of methanol'. All extreme subnetworks of the first model (cf. Table 6.11) and the extreme subnetworks $\mathcal{E}_1$ and $\mathcal{E}_3$ of the second model (cf. Table 6.12) are stable, but non-mixing stable. The non-mixing stability of all existing extreme subnetworks in both models leads by their combination to the occurrence of the Hopf bifurcations in the whole network as shown in Table 4.6 and in Table 4.8.

TABLE 6.11: Computation of mixing stability in the model "calcium oscillations in the cilia of olfactory sensory neurons"

Subsystem	Result	Time
$\mathcal{E}_1$	false	< 1
$\mathcal{E}_2$	false	< 1
$\mathcal{E}_3$	false	< 1

TABLE 6.12: Computation of mixing stability in the model "electro-oxidation of methanol"

Subsystem	Result	Time(s)
$\mathcal{E}_1$	false	< 1
$\mathcal{E}_2$	false	< 1
$\mathcal{E}_3$	false	< 1

Chapter 7

Summary and Outlook

This work is concerned with the investigation of new symbolic methods for detecting the existence of oscillations in complex reaction networks. Hence, The development of new algorithmic approaches for computation of Hopf bifurcation fixed points was the main goal, due to the relation of the occurrence of Hopf bifurcation to the existence of local oscillations. Since the available symbolic algorithmic methods for computing Hopf bifurcations in concentration space turned out to be difficult to handle high-dimensional models, we were motivated to develop new methods, which allow overcoming the difficulties caused by high-dimensionality of the networks and also by the additional conservation laws. Stoichiometric network analysis provides a possibility to simplify the analysis of networks by enabling the decomposition of the whole network and then separably analysis of the individual subnetworks. Using our first method $HoCoQ$, which combines the ideas of stoichiometric analysis, manifold reduction, and quantifier elimination; we could compute Hopf bifurcations in an attempted example involving four species and also in some models, for which the previous available methods fail. However, for some chemical networks with complex dynamics it remained difficult to process the final obtained quantified formulae with the currently available quantifier elimination packages. Therefore we developed a second method for computing Hopf bifurcations called $HoCaT$, which is based on tropical geometry and enables us to refrain from utilizing quantifier elimination. In contrast to $HoCoQ$ the $HoCaT$ method uses a criterion for the occurrence of Hopf bifurcations without requiring empty unstable manifolds. This leads to reduce the problem of existence of Hopf bifurcation fixed points to the algorithmic problem whether a single multivariate polynomial has a zero for positive coordinates. Using $HoCaT$ method we could compute realistic and well-known (bio)-chemical models e.g. mitogen-activated protein kinase (MAPK) and Methylene Blue Oscillator System (MBO). It also works very well for the attempted example involving more than 20 chemical species. We also investigate the existence of Hopf bifurcations in original family of

Chapter 7. Summary and Outlook

models of genetic circuits, for which only the results of quasi-state approximation were available. We conclude that eliminating fast reactions in reaction systems for quasi-steady state approximation may cause the disappearance of Hopf bifurcations.

The pre-processing step of both methods required generation of algebraic data from the (bio)-chemical reaction networks description. Therefore we developed a framework called *(PoCaB)* that, in addition to the integration of the developed algorithmic methods, enables to generate relevant algebraic entities such as stoichiometric matrices and their factorizations, kinetic matrices, extreme currents, polynomial systems, deficiencies and differential equations. We also use *PoCaB* to extract and compute algebraic entities form different biological models obtained from two publicly available databases and we provide the results for public use as large derived database.

The instability in (bio)-chemical systems gives rise to the occurrence of oscillations. So the computation of stability was a further investigated topic in this work. We were able to compute stability in models involving up to 4 chemical species using also the ideas of stoichiometric network analysis, manifold reduction, and quantifier elimination. We studied thereby two conditions for stability, namely *Hurwitz criterion* and *Gantmacher-Stieljes criterion*. We concluded that both criteria yield same results, but the computation of stability using *Hurwitz criterion* are quite faster than using *Gantmacher-Stieljes criterion*. A further term for computing stability namely *mixing stability* was also implemented and tested on various (bio)-chemical networks.

In addition to the computation of Hopf bifurcations and stability, we also investigated the existence of global oscillations. We studied Muldowney's extension of the classical Bendixson-Dulac criterion for excluding periodic orbits to higher dimensions for polynomial vector fields and we discussed the use of simple conservation constraints and the use of parametric constraints for describing simple convex polytopes on which periodic orbits can be excluded by Muldowney's criteria.

Besides enhancing and improving the methods presented in this book we will concern ourselves with the following main points in our further investigations:

1. *PoCaB* provides recently a derived database involving various algebraic data for public use. A main goal in further work is to extend it to an open software for the symbolic analysis of (bio)-chemical reaction networks. We make the existing and newly developed methods accessible to researchers in several fields e.g. computation chemistry and system biology.

2. The efficiency of our method *HoCaT* that can handle even the attempted 22-dimensional model motivated us to apply it on other realistic models and to extend the dimensions of systems that can be handled.

3. In our basic experiment with quasi-steady state approximation we concluded that the occurrence of Hopf bifurcations is also depend on the fast reactions. In further investigations we will analyse the structure of the differential algebraic equations arising in the QSSA. One obvious question concerns the validity of the approximation and another question concerns the effect of the non-linearity of the arising algebraic equations.

4. Investigation of complex differential algebraic equations (DAEs). The simple case of (DAEs) is already investigated in this work and a method for manifold reduction for systems with linear constraints is presented.

5. Investigation of symbolic models for model reduction such a submodel extraction and model transformation.

6. Annotating SBML models with information obtained by the developed methods.

Bibliography

[1] R. Beardmore and K. Webster. A Hopf bifurcation theorem for singular differential algebraic equations. *Mathematics and Computers in Simulation*, 79(4):1383–1395, 2008.

[2] Ivar Bendixson. Sur les curbes définiés par des équations différentielles. *Acta Math.*, 24:1–88, 1901.

[3] Sebastian Bonhoeffer, John M. Coffin, and Martin A. Nowak. Human immunodeficiency virus drug therapy and virus load. *The Journal of Virology*, 71(4):3275, 1997.

[4] François Boulier, Marc. Lefranc, François. Lemaire, Pierre. Morant, and Asli. Ürgüplü. On proving the absence of oscillations in models of genetic circuits. In H. Anai, H. Horimoto, and T. Kutsia, editors, *Algebraic Biology (AB 2007)*, volume 4545 of *Lecture Notes in Computer Science*, pages 66–80. Springer-Verlag, 2007.

[5] François Boulier, Marc Lefranc, François Lemaire, and Pierre-Emmanuel Morant. Applying a rigorous quasi-steady state approximation method for proving the absence of oscillations in models of genetic circuits. In Katsuhisa Horimoto, Georg Regensburger, Markus Rosenkranz, and Hiroshi Yoshida, editors, *Algebraic Biology (AB 2008) – Third International Conference*, volume 5147 of *Lecture Notes in Computer Science*, pages 56–64, Castle of Hagenberg, Austria, August 2008. Springer-Verlag.

[6] Christopher W. Brown. QEPCAD B: A system for computing with semi-algebraic sets via cylindrical algebraic decomposition. *ACM SIGSAM Bulletin*, 38(1):23–24, 2004.

[7] Christopher W. Brown, M'hammed El Kahoui, Dominik Novotni, and Andreas Weber. Algorithmic methods for investigating equilibria in epidemic modeling. *Journal of Symbolic Computation*, 41(11):1157–1173, 2006.

[8] Leonid A. Cherkas. Quadratic systems with maximum number of limit cycles. *Differential Equations*, 45:1440–1450.

[9] Leonid A. Cherkas. Dulac function for polynomial autonomous systems on a plane. *Differential Equations*, 33:692–701, 1997.

[10] Leonid A. Cherkas and Alexander Grin. Algebraic aspects of finding a Dulac function for polynomial autonomous systems on the plane. *Differential Equations*, 37:411–417, 2001.

[11] Leonid A. Cherkas and Alexander Grin. On a Dulac function for the Kukles system. *Differential Equations*, 46:818–826, 2010.

[12] Bruce L. Clarke. *Stability of Complex Reaction Networks*, volume XLIII of *Advances in Chemical Physics*. Wiley Online Library, 1980.

[13] Carsten Conradi, Dietrich Flockerzi, and Jörg Raisch. Multistationarity in the activation of a MAPK: Parametrizing the relevant region in parameter space. *Mathematical Biosciences*, 211(1):105–131, 2008.

[14] Markus W. Covert, Cristophe H. Schilling, and Bernhard Palsson. Regulation of gene expression in flux balance models of metabolism. *Journal of theoretical biology*, 213(1):73–88, November 2001.

[15] Leonardo De Moura and Nikolaj Bjørner. Z3: An efficient SMT solver. In *Tools and Algorithms for the Construction and Analysis of Systems*, pages 337–340. Springer, 2008.

[16] Andreas Dolzmann and Lorenz A. Gilch. Generic Hermitian quantifier elimination. In John A. Campbell Bruno Buchberger, editor, *Artificial Intelligence and Symbolic Computation: 7th International Conference, AISC 2004, Linz, Austria*, volume 3249 of *Lecture Notes in Computer Science*, pages 80–93. Springer-Verlag, Berlin, Heidelberg, 2004.

[17] Andreas Dolzmann and Thomas Sturm. Redlog: Computer algebra meets computer logic. *ACM SIGSAM Bulletin*, 31(2):2–9, June 1997.

[18] Andreas Dolzmann and Thomas Sturm. Simplification of quantifier-free formulae over ordered fields. *Journal of Symbolic Computation*, 24(2):209–231, August 1997.

[19] Mirela Domijan. *Mathematical aspects of chemical reaction networks*. PhD thesis, University of Warwick, 2008.

[20] Mirela Domijan and Markus Kirkilionis. Bistability and oscillations in chemical reaction networks. *Journal of Mathematical Biology*, 59(4):467–501, 2009.

[21] Andreas Dräger, Nicolas Rodriguez, Marine Dumousseau, Alexander Dörr, Clemens Wrzodek, Roland Keller, Sebastian Fröhlich, Nicolas Le Novère, Andreas Zell, and Michael Hucka. JSBML : a flexible and entirely Java-based library for working with SBML. *Bioinformatics*, page 4, 2011.

[22] Henri Dulac. Recherche des cycles limites. *CR Acad. Sci. Paris*, 204:1703–1706, 1937.

[23] M'hammed El Kahoui and Andreas Weber. Deciding Hopf bifurcations by quantifier elimination in a software-component architecture. *Journal of Symbolic Computation*, 30(2):161–179, August 2000.

[24] Michal Feckan. A generalization of Bendixson's criterion. *Proceedings American Mathematical Society*, 129(11):3395–3400, 2001.

[25] Martin Feinberg. Chemical reaction network structure and the stability of complex isothermal reactors. the deficiency zero and deficiency one theorems. *Chemical Engineering Science*, 42(10):2229–2268, 1987.

[26] Akira Funahashi, Mineo Morohashi, Hiroaki Kitano, and Naoki Tanimura. Celldesigner: a process diagram editor for gene-regulatory and biochemical networks. *BIOSILICO*, 1(5):159 – 162, 2003.

[27] Felix R. Gantmacher. *Application of the Theory of Matrices*. Interscience Publishers, New York, 1959.

[28] Fexlix R. Gantmacher. *The theory of matrices. 2*. AMS Chelsea Publishing Series. American Mathematical Society, 2000.

[29] Karin Gatermann. Counting stable solutions of sparse polynomial systems in chemistry. In *Symbolic computation: solving equations in algebra, geometry, and engineering: proceedings of an AMS-IMS-SIAM Joint Summer Research Conference on Symbolic Computation*, volume 286, page 53. American Mathematical Society, 2001.

[30] Karin Gatermann, Markus Eiswirth, and Anke Sensse. Toric ideals and graph theory to analyze Hopf bifurcations in mass action systems. *Journal of Symbolic Computation*, 40(6):1361–1382, 2005.

[31] Karin Gatermann and Birkett Huber. A family of sparse polynomial systems arising in chemical reaction systems. Technical Report Preprint SC 99-27, Konrad-Zuse-Zentrum für Informationstechnik Berlin, 1999.

[32] Karin Gatermann and Birkett Huber. A family of sparse polynomial systems arising in chemical reaction systems. *Journal of Symbolic Computation*, 33(3):275–305, 2002.

[33] Ewgenij Gawrilow and Michael Joswig. Polymake: a framework for analyzing convex polytopes. In Gil Kalai and Günter M. Ziegler, editors, *Polytopes—Combinatorics and Computation*, volume 29 of *Oberwolfach Seminars*, pages 43–73. Birkhäuser Basel, 2000.

[34] Albert Gevorgyan, Mark G Poolman, and David a Fell. Detection of stoichiometric inconsistencies in biomolecular models. *Bioinformatics (Oxford, England)*, 24(19):2245–51, October 2008.

[35] Lorenz A. Gilch. Effiziente Hermitesche Quantorenelimination. Diploma thesis, Universität Passau, D-94030 Passau, Germany, September 2003.

[36] John Guckenheimer and Philip Holmes. *Nonlinear Oscillations, Dynamical Systems, and Bifurcations of Vector Fields*, volume 42 of *Applied Mathematical Sciences*. Springer-Verlag, 1983.

[37] John Guckenheimer and Philip Holmes. *Nonlinear Oscillations, Dynamical Systems, and Bifurcations of Vector Fields*, volume 42 of *Applied Mathematical Sciences*. Springer-Verlag, 1990.

[38] Karl P. Hadeler and Pauline van den Driessche. Backward bifurcation in epidemic control. *Mathematical Biosciences*, 146(1):15–35, 1997.

[39] Hoon Hong, Richard Liska, and Stanly Steinberg. Testing stability by quantifier elimination. *Journal of Symbolic Computation*, 24(2):161–187, August 1997.

[40] Michael Hucka, Andrew Finney, Herbert M Sauro, Hamid Bolouri, John C Doyle, Hiroaki Kitano, Adam P Arkin, Benjamin J Bornstein, Dennis Bray, Athel Cornish-Bowden, et al. The systems biology markup language (SBML): a medium for representation and exchange of biochemical network models. *Bioinformatics*, 19(4):524–531, 2003.

[41] Adolf Hurwitz. Über die Bedingung, unter welchen eine Gleichung nur Wurzeln mit negativen reellen Teilen besitzt. *Math. Ann.*, 46:273–284, 1895.

[42] Yu Ilyashenko. Centennial history of Hilbert's 16th Problem. *Bull. Am. Math. Soc., New Ser.*, 39(3):301–354, 2002.

[43] Abdelhalim Larhlimi. *New Concepts and Tools in Constraint-based Analysis of Metabolic Networks*. PhD thesis, Berlin,Germany, 2008.

[44] Aless Lasaruk and Thomas Sturm. Weak integer quantifier elimination beyond the linear case. In V. G. Ganzha, E. W. Mayr, and E. V. Vorozhtsov, editors, *Computer Algebra in Scientific Computing. Proceedings of the CASC 2007*, volume 4770 of *Lecture Notes in Computer Science*, pages 275–294. Springer, Berlin, Heidelberg, 2007.

[45] Aless Lasaruk and Thomas Sturm. Weak quantifier elimination for the full linear theory of the integers. A uniform generalization of Presburger arithmetic. *Applicable Algebra in Engineering, Communication and Computing*, 18(6):545–574, December 2007.

[46] Xiao-Dong Lin and Pauline van den Driessche. A threshold result for an epidemiological model. *Journal of Mathematical Biology*, 30(6):647–654, 1992.

[47] Wei-Min Liu. Criterion of Hopf bifurcations without using eigenvalues. *Journal of Mathematical Analysis and Applications*, 182(1):250–256, 1994.

[48] Francisco Llaneras and Jesús Picó. Which metabolic pathways generate and characterize the flux space? A comparison among elementary modes, extreme pathways and minimal generators. *Journal of biomedicine & biotechnology*, 2010:753904, January 2010.

[49] Edward N. Lorenz. Deterministic nonperiodic flow. *Journal of the Atmospheric Sciences*, 20(2):130–141, 1963.

[50] Nick I. Markevich, Jan B. Hoek, and Boris N. Kholodenko. Signaling switches and bistability arising from multisite phosphorylation in protein kinase cascades. *Science Signaling*, 164(3):353, 2004.

[51] James S. Muldowney. Compound matrices and ordinary differential equations. *Rocky Mt. J. Math.*, 20(4):857–872, 1990.

[52] Wei Niu and Dongming Wang. Algebraic approaches to stability analysis of biological systems. *Mathematics in Computer Science*, 1(3):507–539, 2008.

[53] Bela Novak and John J Tyson. Modeling the control of DNA replication in fission yeast. *Proceedings of the National Academy of Sciences*, 94(17):9147–9152, 1997.

[54] Luciano Orlando. Sul problema di hurwitz relativo alle parti reali delle radici di un' equazione algebrica. *Mathematische Annalen*, 71(2):233–245, 1911.

[55] Osvaldo Osuna and Gabriel Villasenor. On the Dulac functions. *Qualitative Theory of Dynamical Systems*, pages 1–7, 2011.

[56] Bernhard Palsson. The challenges of in silico biology. *Nature Biotechnology*, 18(11):1147–1150, 2000.

[57] Mercedes Pérez Millán, Alicia Dickenstein, Anne J. Shiu, and Carsten Conradi. Chemical reaction systems with toric steady states. *Bulletin of Mathematical Biology*, pages 1–29, October 2011.

[58] José M. Ponciano and Marcos A. Capistrán. First principles modeling of nonlinear incidence rates in seasonal epidemics. *PLoS Computational Biology*, 7(2):e1001079, 2011.

[59] Brian Porter. *Stability criteria for linear dynamical systems*. Academic Press, New York, 1967.

[60] Patrick J Rabier. The Hopf bifurcation theorem for quasilinear differential-algebraic equations. *Computer methods in applied mechanics and engineering*, 170(3):355–371, 1999.

[61] Patrick J. Rabier and Werner C. Rheinboldt. Theoretical and numerical analysis of differential-algebraic equations. In P. G. Ciarlet and J. L. Lions, editors, *Handbook of Numerical Analysis*, volume VIII, pages 183–540. North-Holland, Amsterdam, 2002.

[62] Richard H. Rand and Dieter Armbruster. *Perturbation Methods, Bifurcation Theory and Computer Algebra*, volume 65 of *Applied Mathematical Sciences*. Springer-Verlag, 1987.

[63] Stefan Ratschan. Efficient solving of quantified inequality constraints over the real numbers. *ACM Transactions on Computational Logic*, 7(4):723–748, 2006.

[64] John Reidl, Peter Borowski, anke Sensse, Jens Starke, Martin Zapotocky, and Markus Eiswirth. Model of calcium oscillations due to negative feedback in olfactory cilia. *Biophysical journal*, 90(4):1147–55, February 2006.

[65] Ricardo Riaza. *Differential-Algebraic Systems*. World Scientific, Hackensack, 2008.

[66] Sonja Sauerbrei, Melke. A Nascimento, Markus Eiswirth, and Hamilton Varela. Mechanism and model of the oscillatory electro-oxidation of methanol. *The Journal of Chemical Physics*, 132(15):154901–154901, 2010.

[67] Christophe H. Schilling, David Letscher, and Bernhard O. Palsson. Theory for the systemic definition of metabolic pathways and their use in interpreting metabolic function from a pathway-oriented perspective. *Journal of Theoretical Biology*, 203(3):229 – 248, 2000.

[68] Alexander Schrijver. *Theory of Linear and Integer Programming*. Wiley Series in Discrete Mathematics & Optimization. John Wiley & Sons, 1998.

[69] Stefan Schuster and Claus Hlgetag. On elementary flux modes in biochemical reaction systems at steady state. *Journal of Biological Systems*, 2(2):165–182, 1994.

[70] Werner M. Seiler. *Involution — The Formal Theory of Differential Equations and its Applications in Computer Algebra.*, volume 24. Springer, 2010.

[71] Anke Sensse. *Convex and toric geometry to analyze complex dynamics in chemical reaction systems*. Phd thesis, Universität Magedeburg, 2005.

[72] Anke Sensse and Markus Eiswirth. Feedback loops for chaos in activator-inhibitor systems. *Journal of Chemical Physics*, 122(4):44516–44700, 2005.

[73] Anne J. Shiu. *Algebraic methods for biochemical reaction network theory*. Phd thesis, University of California, Berkeley, 2010.

[74] Nicola Soranzo and Claudio Altafini. Ernest: a toolbox for chemical reaction network theory. *Bioinformatics*, 25(21):2853–2854, 2009.

[75] Adam W. Strzebonski. Solving systems of strict polynomial inequalities. *Journal of Symbolic Computation*, 29(3):471–480, March 2000.

[76] Adam W. Strzebonski. Cylindrical algebraic decomposition using validated numerics. *J. Symb. Comput.*, 41(9):1021–1038, 2006.

[77] Thomas Sturm. New domains for applied quantifier elimination. In Victor G. Ganzha, Ernst W. Mayr, and E. V. Vorozhtsov, editors, *Computer Algebra in Scientific Computing: 9th International Workshop, CASC 2006, Chisinau, Moldova, September 11-15, 2006*, volume 4194 of *Lecture Notes in Computer Science*. Springer, Berlin, Heidelberg, 2006.

[78] Thomas Sturm. Redlog online resources for applied quantifier elimination. *Acta Academiae Aboensis, Ser. B*, 67(2):177–191, 2007.

[79] Thomas Sturm and Andreas Weber. Investigating generic methods to solve Hopf bifurcation problems in algebraic biology. In Katsuhisa Horimoto, Georg Regensburger, Markus Rosenkranz, and Hiroshi Yoshida, editors, *Algebraic Biology – Third International Conference (AB 2008)*, volume 5147 of *Lecture Notes in Computer Science*, Castle of Hagenberg, Austria, 2008. Springer-Verlag.

[80] Thomas Sturm, Andreas Weber, Essam O. Abdel-Rahman, and M'hammed El Kahoui. Investigating algebraic and logical algorithms to solve Hopf bifurcation

problems in algebraic biology. *Mathematics in Computer Science*, 2(3):493–515, March 2009.

[81] Bernd Sturmfels. *Solving Systems of Polynomial Equations*. American Mathematical Society, Providence, RI, 2002.

[82] Neil Swainston, Kieran Smallbone, Pedro Mendes, Douglas Kell, and Norman Paton. The subliminal toolbox: automating steps in the reconstruction of metabolic networks. *Journal of integrative bioinformatics*, 8(2):186, January 2011.

[83] Alfred Tarski. *A Decision Method for Elementary Algebra and Geometry*. University of California Press, Berkeley, second edition, 1951.

[84] Marco Terzer. *Large Scale Methods to Enumerate Extreme Rays and Elementary Modes*. PhD thesis, 2009.

[85] Janos Toth. Bendixson-type theorems with applications. *Z. Angew. Math. Mech*, 67:31–35, 1987.

[86] Henry C. Tuckwell and Frederic Y. M. Wan. On the behavior of solutions in viral dynamical models. *BioSystems*, 73(3):157–161, 2004.

[87] Robert Urbanczik. Enumerating constrained elementary flux vectors of metabolic networks. *IET Systems Biology*, 1(5):274–279, 2007.

[88] Pauline van den Driessche and Mary L. Zeeman. Three-dimensional competitive Lotka-Volterra systems with no periodic orbits. *SIAM J. Appl. Math.*, 58(1):227–234, 1998.

[89] Vaithianathan Venkatasubramanian, Heinz Schättler, and John Zaborszky. Local bifurcations and feasibility regions in differential-algebraic systems. *Automatic Control, IEEE Transactions on*, 40(12):1992–2013, 1995.

[90] Harald Bernd von Sosen. *Part i: Folds and Bifurcations in the Solutions of Semi-Explicit Differential-Algebraic Equations*. Doctoral dissertation, California Institute of Technology, May 1994.

[91] Clemens Wagner and Robert Urbanczik. The geometry of the flux cone of a metabolic network. *Biophysical Journal*, 89(6):3837–3845, 2005.

[92] Andreas Weber, Thomas Sturm, and Essam O. Abdel-Rahman. Algorithmic global criteria for excluding oscillations. *Bulletin of Mathematical Biology*, 73(4):899–917, 2011.

[93] Andreas Weber, Thomas Sturm, Werner M. Seiler, and Essam Abdel-Rahman O. Parametric qualitative analysis of ordinary differential equations: Computer algebra methods for excluding oscillations. In Vladimir P. Gerdt, Wolfram Koepf, E. W. Mayr, and E. V. Vorozhtsov, editors, *Computer Algebra in Scientific Computing*, volume 6244 of *Lecture Notes in Computer Science*, pages 267–279. Springer-Verlag, September 2010.

[94] Andreas Weber, Martin Weber, and Paul Milligan. Modeling epidemics caused by respiratory syncytial virus (RSV). *Mathematical Biosciences*, 172(2):95–113, September 2001.

[95] Volker Weispfenning. The complexity of linear problems in fields. *Journal of Symbolic Computation*, 5(1&2):3–27, February–April 1988.

[96] Volker Weispfenning. Quantifier elimination for real algebra—the quadratic case and beyond. *Applicable Algebra in Engineering Communication and Computing*, 8(2):85–101, February 1997.

[97] Volker Weispfenning. Quantifier elimination for real algebra—the quadratic case and beyond. *Applicable Algebra in Engineering Communication and Computing*, 8(2):85–101, February 1997.

[98] Volker Weispfenning. A new approach to quantifier elimination for real algebra. In B.F. Caviness and J.R. Johnson, editors, *Quantifier Elimination and Cylindrical Algebraic Decomposition*, Texts and Monographs in Symbolic Computation, pages 376–392. Springer, Wien, New York, 1998.

[99] Clemens Wrzodek, Andreas Dräger, and Andreas Zell. Keggtranslator: visualizing and converting the kegg pathway database to various formats. *Bioinformatics (Oxford, England)*, 27(16):2314–2315, July 2011.

[100] Pei Yu. Closed-form conditions of bifurcation points for general differential equations. *International Journal of Bifurcation and Chaos*, 15(4):1467–1483, April 2005.

[101] Günter M. Ziegler. *Lectures on Polytopes (Graduate Texts in Mathematics)*. Springer, 2001.

I want morebooks!

Buy your books fast and straightforward online - at one of the world's fastest growing online book stores! Environmentally sound due to Print-on-Demand technologies.

Buy your books online at
www.get-morebooks.com

Kaufen Sie Ihre Bücher schnell und unkompliziert online – auf einer der am schnellsten wachsenden Buchhandelsplattformen weltweit! Dank Print-On-Demand umwelt- und ressourcenschonend produziert.

Bücher schneller online kaufen
www.morebooks.de

OmniScriptum Marketing DEU GmbH
Heinrich-Böcking-Str. 6-8
D - 66121 Saarbrücken

Telefax: +49 681 93 81 567-9

info@omniscriptum.de
www.omniscriptum.de

Printed by Books on Demand GmbH, Norderstedt / Germany